DIRECT CURRENT CIRCUIT ANALYSIS
Through Experimentation

Kenneth A. Fiske   and   James H. Harter

Los Angeles Harbor College

4th Edition

The Technical Education Press
Seal Beach, California

DIRECT CURRENT CIRCUIT ANALYSIS
Through Experimentation

ISBN: 0-911908-17-X

Manufactured in The United States of America

Cover Design by Carlos Rico

10 9 8 7 6 5 4 3 2 1

PREFACE

The authors of **DIRECT CURRENT CIRCUIT ANALYSIS THROUGH EXPERI-MENTATION** have long recognized the need for an introduction to circuit analysis for the technicians at the DC resistive circuit level. At first, it was believed that the need could be served by developing an adequate laboratory manual—something away from the cookbook approach—something for the thinking technician. Consequently, the authors proceeded to develop a laboratory manual.

Quickly, it became evident that more than the bare experiments were necessary. If nothing more than the skeletons of experiments were developed, the presentation degenerated back to the cookbook approach. If the manual were keyed directly to available textbooks, the objectives for developing the manual would be lost. Consequently, it became necessary to develop the theory to support the experiments so that topics and ideas could be presented in a definite logical manner. The prime consideration was to maintain student awareness.

Too many students work in the dark when they go into the laboratory. They merely go through the motions with the prime motive of finishing the experiment, rather than the real purpose of gaining depth of understanding. Once the student ceases to understand what he is doing, he loses his interest. This laboratory manual is designed for student involvement so he can acquire confidence and an intuitive "feel" for the subject. Sufficient theory is covered in the introductory portions of the experiment to support the hardware activities of the student as he proceeds through the experiments. The experiments are sequenced in a logical order so a maximum transfer of learning can take place.

The authors are grateful to Faybeth Harter who typed the original manuscript and contributed toward improvements in grammar and style.

Kenneth A. Fiske

James H. Harter

ELECTRONIC CIRCUIT ANALYSIS SERIES
from
The Technical Education Press

DIRECT CURRENT CIRCUIT ANALYSIS
Through Experimentation - 4th Edition
      Kenneth A. Fiske and James H. Harter

ALTERNATING CURRENT CIRCUIT ANALYSIS
Through Experimentation - 3rd Edition
      Kenneth A. Fiske and James H. Harter

BASIC SOLID STATE ELECTRONIC CIRCUIT ANALYSIS
Through Experimentation
      Lorne MacDonald

PRACTICAL ANALYSIS OF AMPLIFIER CIRCUITS
Through Experimentation - 3rd Edition
      Lorne MacDonald

PRACTICAL ANALYSIS OF ELECTRONIC CIRCUITS
Through Experimentation
      Lorne MacDonald

DIGITAL CIRCUIT LOGIC AND DESIGN
Through Experimentation
      Darrell D. Rose

Table of Contents

EXPERIMENT

APPENDIX

1 ANALOG METER FUNDAMENTALS

OBJECTIVES

Basic electrical measurements are made with either a digital or an analog meter. Thus, the knowledge of meters and the ability to extract data from the meter scale is essential to the student working with electricity. This experiment introduces the analog meter with major emphasis placed on the extraction of data from the meter scale.

THE METER

Electronic measuring instruments are used to indicate or record the operation and performance of electric circuits. The most fundamental of these instruments is the analog meter. Figure 1-1 shows the schematic symbol for a meter. A letter or character is placed within the circle to indicate the service the meter is designed to perform.

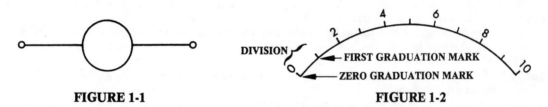

FIGURE 1-1   **FIGURE 1-2**

The most common type of analog meter is a motor mechanism with limited travel. This type of meter consists of a magnetic field, a moving coil with a pointer attached, and a graduated scale. When the meter is properly connected in a circuit, there is a reaction between the magnetic field and the coil which causes the pointer to move over the scale. The distance the pointer travels depends upon circuit conditions which are determined by reading the meter scale.

READING THE METER SCALE

Figure 1-2 shows a meter scale that has been divided into 10 divisions by nine graduation marks. Notice that each division is exactly one tenth of the total distance and represents one tenth of the **full scale value.** Because each division is equally spaced, this scale is called a linear scale.

Figures 1-3(a), (b), and (c) show the same scale with the pointer indicating 4, 7, and 8.5 units, respectively. In Figure 1-3(b), the number 7 has not been printed. However, since the scale is linear, it is obviously the 7th **primary graduation mark.** Figure 1-3(c) shows the pointer halfway between the 8th and 9th graduation marks. This is interpreted to be 8.5 units. Actually, it might be a number very close to 8.5 such as 8.48 or 8.52. Because of the distance between graduations, the eye cannot accurately gauge

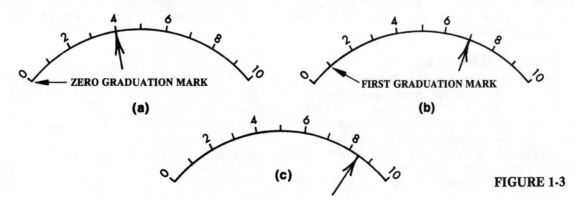

(a)   **(b)**

(c)   **FIGURE 1-3**

the exact distance the pointer has moved. In order to overcome this inadequacy and to improve accuracy, the divisions between primary graduations can be subdivided with additional marks as shown in Figures 1-4 and 1-5.

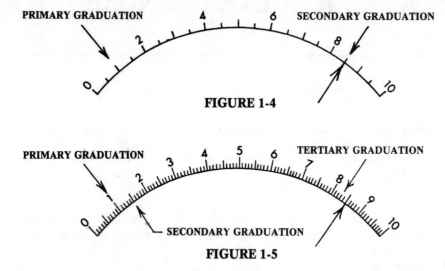

FIGURE 1-4

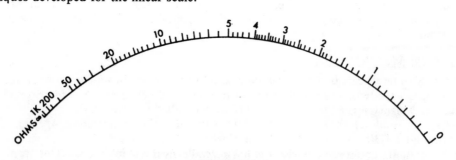

FIGURE 1-5

Both Figure 1-4 and 1-5 have their ten primary divisions subdivided into ten **secondary** divisions. The secondary divisions in Figure 1-5 have been further subdivided into **tertiary** divisions. Now, the pointer in Figure 1-4 clearly indicates 8.5 units. It would still be difficult to accurately state whether or not the pointer is at 8.52 units, but the eye has less distance to subdivide. The ability to gauge more accurately has improved. Further improvement is obtained by increasing the number of subdivisions as shown in Figure 1-5. So, within limits, the accuracy of the **interpolation** increases as the number of secondary and tertiary divisions are increased.

In addition to linear graduations, non-linear graduations are also used quite extensively on meter scales. See Figure 1-6. On the non-linear scale, the distance between the primary graduations diminish as the numerical progression on the scale increases. Data is extracted from a non-linear scale using the same techniques developed for the linear scale.

FIGURE 1-6

NOTE: The number and fineness of graduations on a meter scale depend upon such factors as scale length, meter use, and meter accuracy rating. Good meter reading technique requires careful observation, interpretation, and judgment.

METER TOLERANCE

Another factor that impairs the accuracy of a reading is meter tolerance. In the specification of an instrument, the manufacturer may guarantee a meter reading within plus or minus 3% (±3%) of the full scale deflection. A ±3% specification may not appear to be too detrimental. However, investigation shows that plus or minus 3% of full scale deflection means that 3% of the full scale reading is used to determine the accuracy any place on the scale.

For example, a meter face having 100 units full scale and a ±3% accuracy statement would be plus or minus 3 units any place on the scale. When the pointer is indicating exactly 100 units, the reading is between 97 and 103 units. Now, consider the accuracy when the pointer is deflected halfway, or 50 units.

Since plus or minus 3 units must still be used, the actual reading is somewhere between 47 and 53 units — an error of 6%. The lower on the scale that the pointer indicates, the greater the amount of error.

An equation to determine the percentage of accuracy at any point on a **LINEAR** meter scale is:

$$\frac{FSD \times MSA}{AD} = \text{percentage of accuracy at the deflection point}$$ **EQUATION 1-1**

where: FSD = full scale deflection
MSA = manufacturer's stated accuracy in percent
AD = actual deflection

EXAMPLE 1:

Suppose a meter had a full scale deflection of 300 units and the manufacturer stated the accuracy to be ±2% of full scale. What is the accuracy when the pointer is deflected 200 units?

SOLUTION: Use Equation 1-1.

$$\frac{FSD \times MSA}{AD} = \frac{300 \times 2\%}{200} = 3\%$$

What is the accuracy when the pointer is deflected 125 units?

SOLUTION: Use Equation 1-1.

$$\frac{FSD \times MSA}{AD} = \frac{300 \times 2\%}{125} = 4.8\%$$

EXAMPLE 2:

A certain meter has full scale deflection of 10 units and the manufacturer's stated accuracy is 5%. What is the accuracy when the pointer is deflected 4 units?

SOLUTION: Use Equation 1-1.

$$\frac{FSD \times MSA}{AD} = \frac{10 \times 5\%}{4} = 12.5\%$$

The most accurate or reliable data is obtained from the last two thirds of the linear meter scale. Hopefully, the data could be extracted from the last one half or upper 50% of the scale. Therefore, if it is necessary to read in the lower third of the scale, a more accurate meter should be used because engineering practice requires that meter error not exceed 10%.

THE MULTIMETER

The word "multimeter" refers to a meter that has several ranges and functions. This type of instrument is very popular because of its versatility, compactness, portability, and reliability. The term "function" describes the instrument's versatility in measuring a variety of conditions, and the term "multi-range" describes the capability of the instrument to extend a normally limited range. For each function, then,

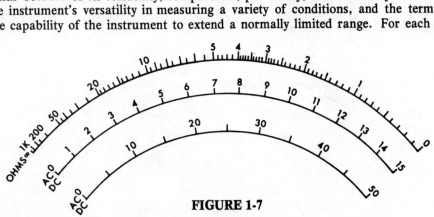

FIGURE 1-7

there are several ranges available. Many manufacturers make multimeters and definite similarities exist between every instrument. The most popular name for the multimeter is the VOM, which are the initials for Volt-Ohm-Milliammeter.

Generally, most VOM meter faces will have one or two non-linear scales and several linear scales. The meter face shown in Figure 1-7 has one non-linear scale and two linear scales. Note that one scale may serve two functions, such as AC and DC. It is customary to place initials or symbols on either or both the right or left hand edges of the scale to identify the function being used. Observe that the meter face of Figure 1-7 has two linear scales marked AC/DC with initials only on the left edge. Note that the numbers 15 and 50 are the upper limits of those scales.

EXAMPLE 3:

Determine which of the two scales of Figure 1-7 to use when the range switch of Figure 1-8 is positioned at 1.5 volts.

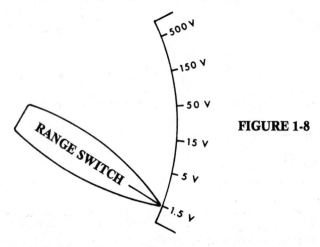

FIGURE 1-8

SOLUTION: The upper scale has a limit of 15, which coincides with 1.5, 15, and 150 volt range positions. By dividing each number on this scale by 10, the full scale reading (upper limit) becomes 1.5. In other words, the 0-15 scale is used, but the full scale deflection is considered to be 0-1.5.

EXAMPLE 4:

Which scale would be used if the range switch is positioned at 500 volts?

SOLUTION: The lower scale. Increase every number on the 0-50 scale by multiplying by 10.

DETERMINING THE ACCURACY OF THE NON-LINEAR SCALE

Having discussed the meter scale and the function and range selector switch or switches, we now return to the procedure for determining the accuracy of the non-linear scale. The simplest method is to first determine the reading on the non-linear scale, and then note the reading on the linear scale. For example, in Figure 1-9, when the pointer indicates a reading of 2.12 units on the non-linear scale, it also crosses 8.0 units on the 0-12 linear scale. Assume the meter of Figure 1-9 has a specified accuracy of 3%. The "**range of the readings**" (tolerance) for the non-linear scale can now be determined.

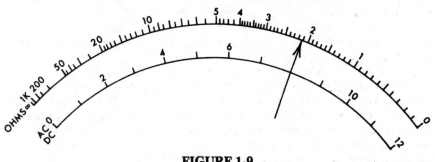

FIGURE 1-9

EXAMPLE 5:

Using a meter with 3% accuracy, determine the range of readings for both the linear and non-linear scales of Figure 1-9.

SOLUTION: Three percent accuracy is ±0.36 units.   Therefore, when the meter indicates 8.0 units, it could be read 7.64 to 8.36 units.

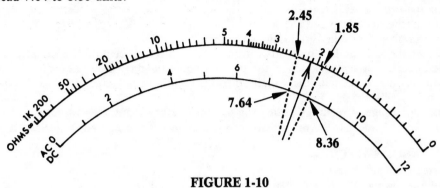

FIGURE 1-10

NOTE: The range of 0.36 units deviation is determined by Equation 1-2.

$$\text{plus or minus range of deviation} = (FSD)(\pm MSA) \qquad \textbf{EQUATION 1-2}$$

SUGGESTED OBSERVATIONS

1.   A meter has a tolerance and, therefore, the meter reading is not absolute. It is, however, an excellent approximation of actual circuit conditions.

2.   A sample "write-up" has been provided in Appendix D, "Preparing the Laboratory Report". It should be consulted before proceeding further.

3.   The tables for data are on separate sheets. Do not write in the tables until the report has been finalized. Use a scratch pad to record rough data.

LIST OF MATERIALS   Alternate Materials

1.   Volt-Ohm-Milliammeter (VOM)
2.   Operator's Manual for Meter
3.   Paper for Computations, Notes

EXPERIMENTAL PROCEDURE

The emphasis in this experiment will be on (a) meter scale interpretations, and (b) proper range selection.

SECTION A: LINEAR SCALES

NOTE: Figure 1-12 (see data pages) shows a multi-range meter scale. To guide you through this section, the authors have followed all of the steps and completed a sample data table.

You are to perform this, and all remaining sections, with the VOM assigned. Record all answers in the space provided in the data table.

1.   What letters or symbols identify the **top-most** linear scale?
2.   Determine the range of the scale.   If this is a multi-range scale, record each range.
3.   How many primary divisions are there for the top range?
4.   How many secondary divisions are there within the top primary division?
5.   Record the unit value of the first primary graduation for each linear scale range.
6.   Record the unit value of the first secondary graduation for each linear scale range.
7.   Record the unit value of the third primary graduation for each linear scale range.
8.   For each scale range, record the unit value of the first secondary graduation mark between the second and third primary graduations.
9.   Using the blank meter scale, sketch in the primary divisions.

10. From the operator's manual or instructor, determine the meter manufacturer's stated accuracy.
11. What is the percentage of accuracy for a one-fourth scale deflection? Use Equation 1-1.

SECTION B: NON-LINEAR SCALES
Place all answers in the data table. (No further examples are provided.)
1. What letters or symbol identifies the uppermost **non-linear** scale?
2. Determine the range of this scale.
3. Is the zero on the right or left of the scale?
4. How many numbered primary graduations are there?
5. Does the number of secondary divisions within primary divisions remain the same?
6. Using the blank meter scale, sketch in the primary and secondary graduations.
7. What is the reading on the non-linear scale when the pointer is deflected 50%? (This would be mid-scale on the linear scales.)
8. Determine the "range of readings" when the pointer is over the number 10 on the non-linear scale. Refer to Example 5 and use the manufacturer's stated accuracy obtained from Step 10 of Section A.

SECTION C: RANGE SWITCH
Place all answers in the data table.
1. Imagine the following DC voltages being measured with your meter:

     (a) 6 V   (b) 0.685 V   (c) 215 V   (d) 75 V

    You are to select and record the best possible DC range available for each listed DC voltage. Record your answers in blank spaces 1(a), 1(b), 1(c), and 1(d). Remember, it is desirable, **whenever possible,** to read all values on the upper two-thirds of the meter scale. However, you may have a meter that cannot provide the readings in the upper two-thirds of the scale.

2. In conjunction with the range switch setting in Step 1, first sketch in the meter face on the blank meter scale of 2(a), 2(b), 2(c), and 2(d) and, then, indicate where the reading will be for 1(a), 1(b), 1(c), and 1(d). (Draw in only the scale divisions that correspond to the range setting.)
3. Does the meter have a separate or a combination range and function switch?

DATA INTERPRETATION AND CONCLUSIONS
    Write a general summary of meter fundamentals presented in this experiment. This discussion should include:
1. Meter accuracy.
2. Linear versus non-linear scales.
3. Function and range switches.
4. Your own conclusions.

APPLICATIONS
    There are numerous applications for the meter, some of which are: (1) automotive fuel, ammeter, and temperature gauges, (2) automobile tachometer, (3) automatic aperture control on your camera, (4) photographic light meter, (5) tape recorder level indicators, and (6) battery charger meter.

PROBLEMS
1. Using Figure 1-11, which range should be selected for a reading of 135 units?

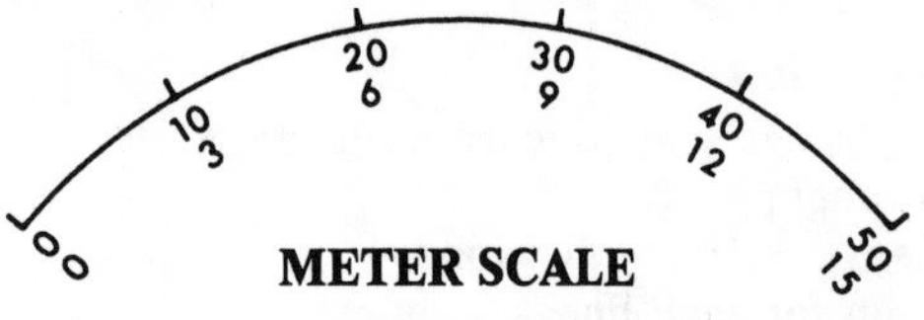

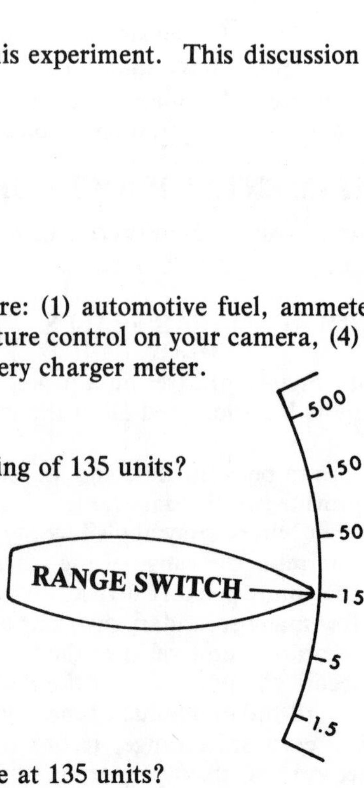

FIGURE 1-11

2. If this meter had a 6% accuracy, what would the accuracy be at 135 units?
3. From a parts catalog, determine the cost of the meter used in this experiment.

2 | RESISTIVE DEVICES AND PARAMETERS

OBJECTIVES

Resistance is basic to the study of electricity and the resistor is one of the most commonly used components in electronic circuits. This experiment introduces you to the unit of resistance, color coding, temperature effects, and wire size. The ohmmeter is introduced and used to make resistance measurements and the value of continuity measurements is stressed.

ELECTRICAL RESISTANCE

Every known substance has the physical property known as electrical resistance. The unit for resistance is the ohm designated by the Greek letter Ω (omega). The schematic symbol for resistance is shown in Figure 2-1. The letter R designates resistance.

FIGURE 2-1

Resistance can be divided into three categories: (1) the conductor—a material that offers a low resistance to the passage of electricity, (2) the semiconductor—a material having a high resistance, but not high enough to be classed as an insulator, and (3) the insulator—a material that has extremely high resistance and offers a great amount of opposition to the flow of electricity.

If an application requires a conductive material, then the selection of a semiconductor or an insulator would be inappropriate. On the other hand, if an application requires a non-conductive material (an insulator) then the other categories would be inappropriate. We see, then, that the application determines whether resistance is a help or a hindrance. The range of **resistivity** for each of the three categories has been established by industry and is available in electrical and electronic handbooks.

Commercial resistors used in electric circuits are classified by (1) construction—i.e. carbon-film, metal-film, or wire wound, (2) wattage rating, and (3) the degree of precision.

RESISTOR VALUES AND THE COLOR CODE

Metal-film and wire wound resistors generally have the value of their resistance printed on the body of the component. Carbon-film and carbon composition resistors are color coded with bands at one end indicating the value of resistance. Table 2-1 lists a standard 4-band, color-code chart used to decode resistor values.

TABLE 2-1: COLOR CODE CHART

COLOR	DIGIT	MULTIPLIER	TOLERANCE
Black	0	$10^0 = 1$	
Brown	1	$10^1 = 10$	
Red	2	$10^2 = 100$	
Orange	3	$10^3 = 1,000$	
Yellow	4	$10^4 = 10,000$	
Green	5	$10^5 = 100,000$	
Blue	6	$10^6 = 1,000,000$	
Violet	7	$10^7 = 10,000,000$	
Gray	8	$10^8 = 100,000,000$	
White	9	$10^9 = 1,000,000,000$	
Gold		$10^{-1} = 0.1$	5 %
Silver		$10^{-2} = 0.01$	10 %
No Color			20 %

Figures 2-2 and 2-3 show resistors with three and four color bands. To decode these resistors, start with the band closest to the end and work toward the center. Thus:

(a) first band = first digit
(b) second band = second digit
(c) third band = multiplier
(d) fourth band = tolerance
    (or no color)

NOTE: An extremely wide "first band" indicates a wire wound resistor. Also be aware that a five band color code system is also in use.

Figure 2-2(b) shows an example of three band coding. Notice that the first band is orange (3), the second band is white (9), and the third band is yellow. The third band is a multiplier of 10,000. Thus, the resistor has a value of 390,000 ohms or 390 kΩ. (Recall that kilo = k = 1000 and mega = M = 1,000,000.) Since there is no fourth band, the tolerance is ±20% of the coded value.

Figure 2-2(c) is somewhat different in that the third band is gold. Since this band is still a multiplier, then 47 is multiplied by 0.1 to get 4.7 Ω ±20%.

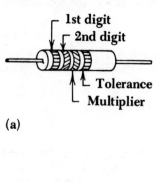

(a)

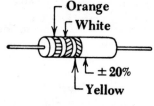

(b) 390,000 Ω ± 20%

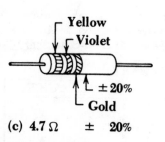

(c) 4.7 Ω ± 20%

FIGURE 2-2

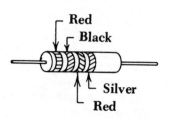

(a) 2,000 Ω ± 10%

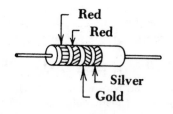

(b) 2.2 Ω ± 10%

FIGURE 2-3

RESISTOR TOLERANCE

When the manufacturer coded his resistor to be 390 kΩ ±20%, he was, in effect, stating that if the resistor were measured and found to have a resistance in the range of 312 kΩ to 468 kΩ that the resistor would be within tolerance. An equation for determining the percentage of difference between the coded value and the measured value is as follows:

$$\% \text{ difference} = \frac{MV - CCV}{CCV} \times 100 \qquad \text{where: } \begin{array}{l} CCV = \text{color coded value} \\ MV = \text{measured value} \end{array} \qquad \textbf{EQUATION 2-1}$$

EXAMPLE 1:

A 12 kΩ 10% resistor has just been purchased (color coded as brown-red-orange-silver). Upon measuring the resistor, it is discovered that it only reads 11.2 kΩ. Is this value within tolerance?

SOLUTION: Ten percent of 12 kΩ is 1.2 kΩ and, therefore, the range is 10.8 kΩ to 13.2 kΩ. So the resistor is within tolerance.

What is the actual percentage of difference?

SOLUTION: Use Equation 2-1.

$$\% \text{ difference} = \frac{MV - CCV}{CCV} \times 100 = \frac{11.2 \text{ k}\Omega - 12 \text{ k}\Omega}{12 \text{ k}\Omega} \times 100 = -6.67\%$$

NOTE: Both new and old resistors should be checked prior to their installation, because the resistance of all resistors change with time, temperature, and usage.

RESISTOR WATTAGE RATING

All resistors have a power or wattage rating. This rating is predetermined by the manufacturer, and it is directly related to the physical size of the resistor. Generally, the larger the resistor, the larger the wattage rating. Since this experiment is concerned only with wattage versus size, further discussion of the wattage concept will be continued in other experiments.

WIRE SIZE AND RESISTANCE

Since some resistors are constructed with wire, the resistive effects of wire must be considered. Wire has resistance which varies with size, length, material, and temperature. Table 2-2 is a condensed wire chart containing selected wire sizes and cross-sectional areas. Remember that the American Wire Gauge (AWG) is the standard that determines both the wire size and area. Consult a handbook for the complete set of wire tables.

The equation for determining the resistance of wire is:

$$R = \rho \frac{L}{A}$$
EQUATION 2-2

where: ρ = rho = the specific resistance in ohms (cmil/ft)
L = length of the wire in feet
A = cross-section in circular mils
R = the resistance of the wire in ohms

This equation states that for any given material, the resistance will increase if the length is increased, and decrease if the cross-section is increased. Table 2-3 contains a few specific resistance numbers (rho) for several selected materials.

GAUGE NUMBER	CIRCULAR MILLS*
10	10,400
12	6,530
14	4,110
16	2,580
18	1,620
20	1,020
26	254

*Some values have been rounded for ease of calculation.

TABLE 2-2: WIRE GAUGE CHART

MATERIAL	Ohms cmil/ft at 20° C (rho)
Copper	10.37
Aluminum	17.00
Tungsten	33.00
Nichrome I	661.00

TABLE 2-3: SPECIFIC RESISTANCE

EXAMPLE 2:

What is the resistance of 100 feet of No. 10 copper wire at 20°C?

SOLUTION: From Table 2-2 it can be seen that No. 10 wire has a cross-section of 10,400 cmil and from Table 2-3 that the rho value for copper is 10.37. Substituting these values into Equation 2-2 results in a resistance of:

$$R = \rho \frac{L}{A} = \frac{10.37 \ \Omega \times (\text{cmil/ft}) \times 100 \ \text{ft}}{10,400 \ \text{cmil}} = 0.10 \text{ ohm}$$

EXAMPLE 3:

A technician has been assigned the responsibility of installing a two-conductor aluminum cable to a water pump used to cool a high-power transmitter tube. This pump is located at a distance of 400 feet from the power service entrance. The resistance of the wire must not exceed 2.5 ohms at an ambient temperature of 20°C. Which of the two conductors, No. 12 or No. 14, should be selected?

SOLUTION: Rho is 17 for aluminum and the cross-sectional areas of No. 12 and No. 14 wire are 6,530 cmil and 4,110 cmil, respectively. The total length of the wire required is 800 feet (the 400 foot run from the service entrance to the pump and the 400 foot return).

(a) The resistance of the No. 12 wire is 2.08 ohms from:

$$R = \rho \frac{L}{A} = \frac{17 \ \Omega \times 800}{6530} = 2.08 \ \Omega$$

(b) The resistance of the No. 14 wire is 3.31 ohms from:

$$R = \rho \frac{L}{A} = \frac{17 \ \Omega \times 800}{4110} = 3.31 \ \Omega$$

Therefore, No. 12 wire should be selected because it has a resistance less than the specified 2.5 ohms.

THE EFFECT OF TEMPERATURE ON THE RESISTANCE OF WIRE

Notice that in both Examples 2 and 3 the temperature must be 20°C, because rho, and hence the temperature coefficient alpha, was tabulated at this temperature. See Tables 2-3 and 2-4. A selection of any other temperature would have invalidated the results of Equation 2-2, because temperatures other than 20°C need to be specified. Therefore, a new equation, Equation 2-3, is needed.

$$\frac{R_1}{R_2} = \frac{[(1/a_t) - t] + T_1}{[(1/a_t) - t] + T_2}$$

EQUATION 2-3

where: R_1 = resistance at T_1 (see Example 6)
R_2 = resistance at T_2 (see Example 6)
a_t = temperature coefficient
t = temperature at which temperature coefficient was stipulated
T_1 = temperature of R_1 in °C
T_2 = temperature of R_2 in °C

TABLE 2-4: TEMPERATURE COEFFICIENTS (Alpha = a_t)

MATERIAL	0°C	20°C
Copper	0.004264	0.00393
Aluminum	0.00423	0.0039
Tungsten	0.0049	0.00446
Nichrome	0.0004032	0.0004

16

Equation 2-3 was derived using the inferred zero resistance concept. A more comprehensive explanation of this concept can be found in several of the references listed in Appendix D-2.

To establish the inferred zero point, just the term $[(1/a_t) - t]$ from Equation 2-3 is required. Its validity will be shown in Examples 4 and 5.

EXAMPLE 4

A handbook states that the temperature coefficient of annealed copper wire is 0.003929 at 20°C. This may be noted as $a_t = 0.003929_{20}$. What is the inferred zero resistance point?

SOLUTION: $[(1/a_t) - t] = [(1/0.003929) - 20] = 234.5°C$

Thus, the inferred resistance point occurs 234.5° C to the left of the origin. See Figure 2-4.

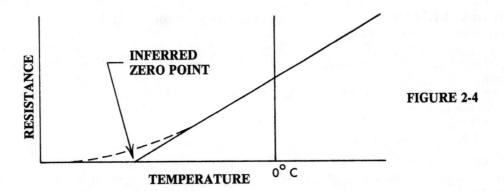

FIGURE 2-4

EXAMPLE 5:

Another handbook shows the temperature coefficient (alpha) as 0.0042644 at 0° C for annealed copper wire. Which of the two coefficients (0.003929 or 0.0042644) are correct?

SOLUTION: $[(1/a_t) - t] = [(1/0.0042644) - 0] = 234.5 - 0 = 234.5°C$

Obviously, both temperature coefficients are correct, since both examples result in the same inferred zero resistance point. Since the same material was used in both examples, the points in questions had to be the same. You should note that alpha is a temperature dependent variable. Therefore, the difference in alphas occurred because of the difference in temperatures (0° C and 20° C).

Example 6 gives more complete insight into the flexibility of Equation 2-3.

EXAMPLE 6:

A certain meter manufacturer states that the coil resistance of a particular meter movement is 2000 ohms at 25° C. The coil was constructed using annealed copper wire. Several hundred of these meters are to be sold to a firm manufacturing test equipment. Assuming that the meter resistance is quite important in determining the accuracy of the test equipment, what will be the change in coil resistance if the instrument is to work in an environment of −55° C to +105° C?

SOLUTION: For the temperature of −55° C, the resulting resistance is 1383 ohms from:

$$\frac{R_1}{R_2} = \frac{[(1/a_t) - t] + T_1}{[(1/a_t) - t] + T_2} = \frac{2000}{R_2} = \frac{[(1/0.00393) - 20] + 25}{[(1/0.00393) - 20] - 55} = \frac{234.5 + 25}{234.5 - 55}$$

$$R_2 = \frac{2000(179.5)}{259.5} = \frac{359000}{259.5} = 1383 \text{ ohms}$$

For the +105° C temperature, the resulting resistance is 2617 ohms from:

$$R_2 = \frac{2000(234.5 + 105)}{259.5} = \frac{679000}{259.5} = 2617 \text{ ohms}$$

17

Thus, it can be seen that the resistance changes from 2000 ohms at 25° C to 1383 ohms at −55° C and to 2617 ohms at +105° C.

SUGGESTED OBSERVATIONS

1. Do not put absolute faith in the color code. Resistors can be mislabeled and resistance can change in value.

2. Do not measure the resistance of an analog meter with an ohmmeter. The destruction of the meter is the usual result.

3. Do not make resistance measurements on components that have power applied to them. Severe damage to the ohmmeter may occur.

4. For proper operation of the ohmmeter, consult the operator's manual or the instructor.

5. Record all measurements and calculations in the data table.

LIST OF MATERIALS

Alternate Materials

1. VOM
2. VOM Operator's Manual
3. 12 Assorted Color Coded Resistors with Assorted Wattage Ratings
4. 6 Wire Wound or Deposited Film Resistors
5. 2 Specially Constructed Test Cables

EXPERIMENTAL PROCEDURE

The emphasis in this experiment is on: (a) measuring and decoding resistors; computing tolerances; evaluating resistance changes due to temperature, wire size, and length; and (b) trouble shooting and continuity checks.

SECTION A: OHMMETER OPERATION

In this section the student will learn to set up the ohmmeter and make resistance measurements. Once the procedure has been established, it will be basic to most ohmmeters. Remember to record the answers in the data tables. Refer back to Experiment 1, notes 2 and 3 or the Suggested Observations section.

1. Insert the test leads in the jacks used for resistance measurements. Consult the operator's manual or the instructor.

2. If the multimeter has separate range and function switches, set the function switch to the position used for resistance measurements as indicated in the operator's manual.

3. Set the range switch to R × 1. With the range switch in position, the ohmmeter scale is read directly. As an example, if the pointer indicated a deflection of 15 units on the ohms scale, then: 15 × 1 = 15 ohms.

4. Connect the test probes together. This process is commonly called "shorting the leads together." The meter should show a deflection. If it does not, the ohmmeter may be set up improperly. Now, adjust the control marked zero ohms, ohms adjust, etc. so that the pointer is over the zero mark of the ohms scale.

NOTE: Test leads should not be left shorted together for long periods of time as it will cause the instrument's battery to discharge.

5. Set the range switch to the next range. This might be R × 15 or R × 10, etc. As an example, assume a pointer deflection of 25 units and that the range selector is on the R × 10 range; the resistance would be 25 × 10 or 250 ohms. If the range selector is on the R × 100 range, the reading would be 2500 ohms or 2.5 kΩ.

6. Every time the range switch is changed, the meter must be rezeroed as outlined in Step 4.

7. Assume that the meter has the following ranges: R × 1, R × 10, R × 1000, and R × 100,000. Record the answers in the data table for a deflection of:
   (a) 12 units on the R × 1 range.
   (b) 5 units on the R × 10 range.
   (c) 50 units on the R × 1000 range.
   (d) 36 units on the R × 100,000 range.

8. Does the ohmmeter scale have a zero left or zero right scale?

9. After having zeroed the meter, connect the test leads to one of the six wire wound or deposited film

resistors.

NOTE: Remember, because of meter accuracy, scale non-linearity, and limited range selection, you will have to decide from which scale you can obtain the best reading.

> Record in column (a) the range setting which gives the most accurate reading.
> Record in column (b) the measured value of the resistor.
> Record in column (c) the value of the resistance stamped on the resistor.

10. Repeat Step 9 for the remaining 5 resistors.

SECTION B: COLOR CODE DECODING AND MEASUREMENTS

1. Using the same techniques developed in Section A, select one of the 12 color coded resistors with axial leads (wires coming out of the ends) and measure its resistance. Record the answers in the data table.
2. Using the same resistor, determine its value by decoding the color bands. Refer to Table 2-1.
   (a) Record the decoded value.
   (b) Is the measured value within the coded tolerance limits?
   (c) Using Equation 2-1, determine the percentage difference between the coded and measured values.
3. Repeat Steps 1 and 2 for all 12 color coded resistors.

SECTION C: CONTINUITY MEASUREMENTS

Two specially constructed cables will be required in this section. Record all measurements and calculations in the data tables.

1. A multi-conductor cable connecting a remote read-out station with a central computer has been severed and spliced back together again. As a precautionary measure, a continuity check is made on the cable after it has been spliced. Using Test Cable No. 1 (single-banded end) and an ohmmeter, determine which of the wires at the banded end have continuity with the wires at the unbanded end. Hint: The wire splicer may have insulated some of the splices improperly. So check for short circuits (continuity) between the wires.
2. A telephone line repairman is at a substation in a mountainous area when the service is suddenly disrupted. Theorizing that a stray bullet has lodged itself within the cable and shorted the wires together, he makes the following test:
   (a) Using his VOM, he measures the resistance between the orange and blue wires of the test cable No. 2 of the banded end. A jumper wire must be connected between the green and blue wires of the unbanded end to represent the short caused by the bullet. Record the measured resistance in the data table.
   (b) Knowing that the cable contains several No. 26 copper wires, he then calculates the estimated distance to the short using Equation 2-4, which is simply a modification of Equation 2-2. Record this estimate in the data table.

$$L = \frac{RA}{\rho \times 5280 \times 2}$$ **EQUATION 2-4**

where   L = length in miles
R = resistance of the wire as measured in Part (a)
A = cross-sectional area in circular mils
ρ = specific resistance = 10.37 Ω (Assume a temperature of 20 degrees C.)
5280 = number of feet in a mile
2 = multiplier used to include the return length of wire   Equation 2-4

   (c) The line repairman now drives toward the estimated point of difficulty. Before reaching his destination, he stops at a junction box and takes another resistance reading. Using either Equation 2-2 or 2-4, determine how far he is from the short by measuring the resistance between the red and brown wires (banded end—Cable No. 2). Remove the jumper in Part (b) and then connect a jumper wire between the red and violet wires—unbanded end. Compute and record this new distance.
3. Draw a schematic diagram of Test Cable No. 2. (How is it constructed?) As shown in Example 7, draw a straight line if there is continuity and no resistance. If there is resistance, use the resistance symbol and mark the value of the resistance next to the symbol. A format for this schematic, with wire

colors, is provided in the data tables.

EXAMPLE 7:

If zero resistance were measured when checking between a green wire and a brown wire, this would be indicated by:

GREEN O————————————————————————O BROWN

A measurement between a blue and yellow wire having a resistance of 350 ohms would be indicated by:

BLUE O————————⋀⋀⋀————————O YELLOW
350 Ω

A measurement between a red and orange wire having an indication of infinity would be indicated by:

RED O————————O   O————————O ORANGE

DATA INTERPRETATION AND CONCLUSIONS

Write a general summary of the ideas presented in this experiment. This discussion should include:
1. Resistor tolerances
2. Wire sizes
3. Temperature
4. Wattage versus size
5. Your own conclusions

APPLICATIONS

Some of the applications where resistance serves a useful purpose are as follows: (1) meter multipliers and shunts, (2) solid state circuitry, (3) strain gauges, (4) photoresistive cells, (5) resistance thermometers, (6) electric blankets, (7) telephones, (8) automobile radio noise suppressors, (9) electric stoves, and (10) voltage dropping resistors.

PROBLEMS

1. Determine the wattage rating of the 12 color-coded resistors used in this experiment by comparing the dimensions of each with those listed in a parts catalog.
2. Determine the price of each resistor.
3. What is the approximate cost of a 100 foot spool of No. 20 stranded wire with vinyl insulation?
4. Having determined the price for 100 feet of No. 20 stranded wire with vinyl insulation, how much cheaper per 100 feet would it be to buy a 1000 foot spool of the same wire?

3 | BATTERIES AND PHOTOVOLTAIC CELLS

OBJECTIVES

All electrical and electronic equipment require a source of electrical energy for operation. This experiment introduces the battery and the photovoltaic cell as two basic electrical energy sources. Theoretical and practical battery symbols are used to develop awareness of hidden (**internal**) resistances of circuit components and devices. Voltage notation is introduced in conjunction with voltage measurements to start you thinking in terms of voltage evaluation in circuit analysis.

THE BATTERY

The battery is one of a variety of sources for electrical energy. A "perfect battery" symbol is shown in Figure 3-1. Batteries are generally categorized as those that can be recharged and reused and those that cannot.

A battery, such as those commonly used in flashlights, cannot be recharged and is called a **primary** or **dry cell**. A carbon rod in the center of the cell serves as the positive electrode and the zinc container serves as the negative electrode. Encased between the two electrodes is a specially compounded wet paste called electrolyte. A chemical reaction between the two electrodes and the electrolyte produces electrical energy.

The **secondary** or **wet cell** is so constructed that the chemical reaction between electrodes and electrolyte can be reversed and the battery recharged. This is accomplished by applying an external voltage source to the discharged battery to restore the electrodes and the electrolyte to their original conditions. The automobile storage battery is a common example of a secondary cell that can be recharged in this manner.

Even though the battery is a source of electrical energy, the electrolyte, like all matter, has resistance. This resistance is present in both the charged (low resistance) and discharged (high resistance) states. The exact nature of this resistance is not easily defined since it encompasses many factors. Nevertheless, this resistance is a tangible property and can be measured. A measuring technique will be developed in a subsequent experiment.

Since resistance is an integral part of the energy (voltage) source, Figure 3-1 will be modified to include the resistance symbol as shown in Figure 3-2. It is this figure that must be used in DC circuit analysis. Other voltage sources must also be shown with this resistance symbol. Example 1 will serve to introduce some of the characteristics needed to predict the type of service a battery can provide.

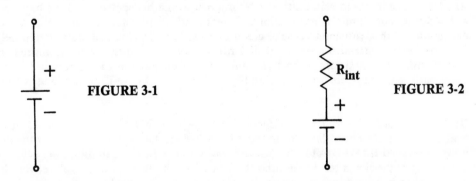

FIGURE 3-1 R_{int} FIGURE 3-2

EXAMPLE 1:

A certain transistor radio manufacturer needs a 9-volt battery to power his radios. The battery manufacturer published a service life curve, Figure 3-3, from which the radio manufacturer can determine how long the battery will last in his radio. It was determined that the radio would cease to function when the battery voltage is reduced to 6 volts. The current drain on the battery is 5 mA resulting from a resistive load of 180 ohms. How many hours of service will the battery provide, if the radio is played 4 hours per day?

SOLUTION: Enter the graph of Figure 3-3 from the voltage axis (see dotted line) at 6 volts and move to the right until the curve of the 180 ohm load is intersected. Now start down toward the hours of service axis. Note that for the conditions stated, 20 hours is the limit of the battery life. In other words, this radio could play for 4 hours per day for 5 days before a new battery is needed.

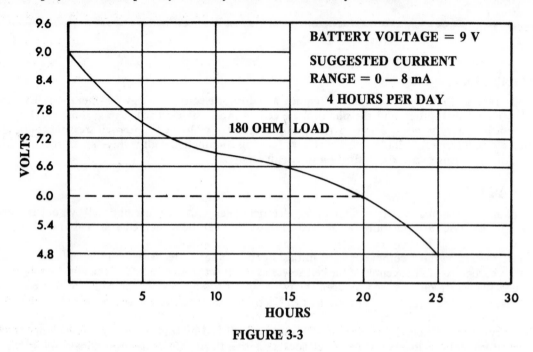

FIGURE 3-3

Batteries are generally classified by their construction. If the battery used in the radio is constructed with zinc and carbon electrodes, it is classed as a "zinc carbon" cell. Other types of batteries include the lead acid cell, the nickel cadmium cell, and the mercury cell. These latter three kinds of cells are rated in **ampere-hour** capacities. This ampere-hour capacity rating is just another way of expressing the capabilities of the battery. Hence, a 10 ampere-hour (Ah) battery can, theoretically, deliver 10 amperes for 1 hour, 2 amperes for 5 hours, or 1 ampere for 10 hours. However, in reality, batteries have a limited current capacity. It is imperative, therefore, that the manufacturer's suggested current range not be exceeded.

Both the lead acid and the nickel cadmium cells can be recharged. The problems involving the recharging of batteries are presented in more advanced courses. However, two points should be noted: (1) the state of charge in a lead acid battery is determined with a **hydrometer** and the battery must never be left in a discharged condition for more than 24 hours, and (2) the state of charge of a nickel cadmium cell is determined with a voltmeter and the cell may be left in a discharged state indefinitely.

Many battery manufacturers publish battery manuals. These give a comprehensive coverage of the manufacture, use, and care of batteries and are available at a nominal cost. You should purchase one of these battery manuals for your working library, so you will have it available as a reference when working with batteries.

NOTE: A word of caution is in order for both the lead acid and mercury cells. It is possible to spill the acid of the lead acid cell. Should this acid be spilled on the clothing or parts of the body, the affected area must be washed IMMEDIATELY with either water or soda water to dilute the acid. The mercury cell has a tendency to develop a residue around the seals as the battery ages. Therefore, your hands must always be washed after handling such a battery. The residue is not harmful to your hands, but it can cause serious illness if taken internally.

THE PHOTOVOLTAIC CELL

Another source of voltage commonly used in satellites and space probes is the photovoltaic or solar cell. These cells give off electrical energy when exposed to light or similar radiation. The symbol for either type of cell is shown in Figure 3-4. Remember that a resistive symbol must be shown as an integral part of the

22

cell. See Figure 3-5.

The output voltage of the photovoltaic cell varies from 0.25 volts for selenium to 0.4 volts for silicon. This voltage is fairly constant for any size cell. The output current of a photovoltaic cell can only be obtained from curves published by the manufacturer or by experimentation. Some of the factors that must be considered when trying to determine the cells' output characteristics are: (1) the type of illumination—infrared, sunlight, incandescent lamp, ultraviolet, etc., (2) the distance from the source of light to the cell, and (3) the intensity of the light source. As a voltage source, the solar cell will have both a positive and negative terminal.

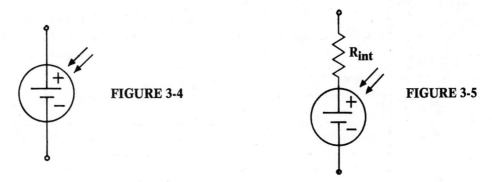

FIGURE 3-4   FIGURE 3-5

VOLTAGE POLARITY AND VOLTAGE NOTATION

The identification of the positive and negative terminals of any voltage source must be made prior to connecting the source to a circuit. In addition, a knowledge of terminal polarity is imperative when analyzing circuit conditions. A simple voltage measurement is used to determine the positive and negative terminals.

A question that arises in regard to polarity is: "positive with respect to what?" To answer this question, recall an earlier statement that the center terminal of a dry cell is positive and the case is negative. This seemingly innocent statement can certainly cause difficulties when trying to analyze circuit conditions. For instance, when examining the circuit of Figure 3-6, note that the voltmeter leads are connected in such a manner that the positive lead of the voltmeter is attached to the positive terminal of the battery, and the negative or common lead is attached to the negative terminal. The voltmeter will, therefore, deflect up scale and indicate a voltage of 1.5 volts. Since the voltmeter will only deflect up scale when connected as shown in Figure 3-6, it can be concluded that the center terminal is positive with respect to the case terminal.

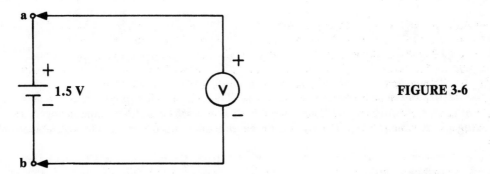

FIGURE 3-6

The letters at each end of the voltage source in Figure 3-6 are used to formulate the symbolic command, E_{ab} or E_{ba}. Using the command E_{ab} means that the voltage is to be determined by using Point **b** as the reference point. In statement form, **the negative lead of the voltmeter must be placed on the terminal designed as b, and the positive lead on terminal a.**

It is extremely important that symbolic commands be complied with. That is, when told to determine E_{ba} for the circuit of Figure 3-6, the procedure would be to connect the negative lead of the voltmeter to the positive battery terminal (Point **a**) and the positive lead to the negative battery terminal (Point **b**). Granted the meter will not function as it was intended, but you complied with the symbolic command. (The meter will deflect down scale when connected like this; however, NO meter damage will result.)

Whether discussing E_{ac} or E_{bd} or any other similar notation in a circuit, remember that the second subscript is the point of reference; and as such, the negative (common) lead of the meter is placed on it. Now the question is, what should be done about the meter reading down scale? Obviously, one cannot tell what the magnitude is unless the meter is deflected up-scale. Therefore, if commanded by notation to obtain the voltage E_{ba} in Figure 3-6, the meter would be connected as commanded. Note that the meter reads down scale, and **then** reverse the meter leads and read the magnitude of the voltage from the meter. Because of the meter lead reversal, a negative sign is placed in front of the answer, i.e. -1.5 volts.

EXAMPLE 2:

In the circuit of Figure 3-7(a), determine E_{ba}.

SOLUTION: $E_{ba} = 3.0$ volts

EXAMPLE 3:

In the circuit of Figure 3-7(a), determine E_{ab}.

SOLUTION: $E_{ab} = -(3.0 \text{ volts})$

The term within the parenthesis is actually E_{ab}. Therefore, it can further be stated that:

$$E_{ab} = -E_{ba}$$

EXAMPLE 4:

Determine E_{cd} in Figure 3-7(b).

SOLUTION: $E_{cd} = 4.5$ volts

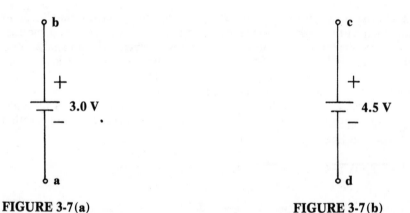

FIGURE 3-7(a)   **FIGURE 3-7(b)**

SUGGESTED OBSERVATIONS

1. An excellent source of information about the VOM is the VOM operator's manual.
2. When taking measurements, handle only one probe at a time and do not touch the metal tip of the probe.
3. Handle the materials very carefully. Keep them away from the edge of the bench as they can be knocked to the floor and damaged.
4. Prior to applying voltage, make certain that the meter has been set up and connected properly.
5. Record all readings and calculations in the data table.

LIST OF MATERIALS

1. VOM
2. VOM Operator's Manual
3. Four 1.5 volt Batteries
4. Photovoltaic Cell
5. 12-inch Ruler
6. 75-100 watt Light Source

Alternate Materials

_____

_____

_____

LIST OF MATERIALS (CONTINUED)

7. Resistors (all at least ½ watt)

4.7 Ω	10 Ω	100 Ω	470 Ω
1 kΩ	1.5 kΩ	2.2 kΩ	2.7 kΩ
3.9 kΩ	4.7 kΩ	5.6 kΩ	6.8 kΩ
10 kΩ			

8. Graph Paper
   3 Sheets of 10 × 10 to the inch
   2 Sheets of semi-log 4 cycles

EXPERIMENTAL PROCEDURE:

The following experiment will show how batteries react when connected in different configurations, and how photovoltaic cells respond to the intensity of the stimulus. In addition, symbolic notation will become quite meaningful.

SECTION A: VOLTAGE MEASUREMENT

Record all readings and calculations in the data tables.

1. Set the VOM function switch to DC volts and the range switch to a value of at least 1.5 volts. (This might be 1.5 volts, 2.5 volts, 3 volts, or some similar range setting.)
2. Insert the test leads in the proper jacks for measuring DC voltages. The black lead is the common test lead, and it is placed in the jack labeled common or negative. Consult the operator's manual or check with the instructor for a detailed set-up procedure.
3. Identify the negative and positive terminals of a 1.5 volt battery.
4. Attach the common (black) lead of the VOM to the negative terminal of the battery.
5. Touch the test probe to the positive terminal of the battery. If the meter pointer deflects down-scale check your test lead placement. If the pointer is deflected beyond the full scale value, **remove** the probe and increase the range setting.
6. Read the appropriate DC scale of the meter. Remember that accurate readings are taken from the middle and upper 1/3 of the meter scale. If the pointer is deflected into the lower 1/3 of the scale, then, it may be possible to select a different range. To see if this is possible, determine if the meter indicates a value that is less than the value of the next lower range setting. If it does, switch to the next lower range.
7. If difficulty is encountered in reading the meter, Experiment 1 should be reviewed.
8. Begin developing safe work habits by using only one hand and one test lead. The use of two hands may result in severe Electrical Shock.

SECTION B: AIDING AND OPPOSING VOLTAGE SOURCES

Record all readings and calculations in the data table.

1. Connect two cells in series aiding. That is, connect the negative terminal of one battery to the positive terminal of the other. If connected properly, two terminals—one positive and one negative—will remain. Using the technique developed in Section A, measure the voltage across both the series connected cells. Make certain that the range selector is set to an appropriate voltage range.
2. Connect two cells in series opposing. This means connect the negative terminal of one cell to the negative terminal of the other. If connected properly, two terminals—both positive—will remain. Measure the voltage across this combination.
3. Connect four cells in series aiding, and measure the voltage across them. Set the range switch to at least 10 volts DC.
4. Connect three cells in series aiding and one cell in series opposing. See Figure 3-8. Measure the voltage across terminals a and b (E_{ba}).

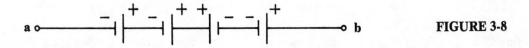

FIGURE 3-8

5. Parallel two cells by connecting them so identical terminals are together. That is, positive to positive and negative to negative. See Figure 3-9(a). **It is very important that paralleled cells not be connected**

positive to negative in a closed loop as in Figure 3-9(b); this configuration will damage the batteries. Measure the voltage across the terminals a and b of Figure 3-9(E_{ab}).

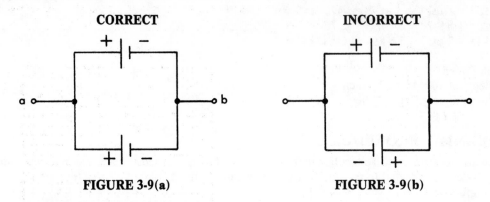

CORRECT

FIGURE 3-9(a)

INCORRECT

FIGURE 3-9(b)

6. Now, parallel three cells and measure this voltage.

SECTION C: INTERNAL RESISTANCE

Suppose a transistor radio is powered with the battery shown in Figure 3-10. As time passes, the internal resistance, R_{int}, of the battery increases until the voltage, as measured at the battery terminals, is reduced to 0.9 volts. This low voltage causes the sound from the radio to become distorted. Assuming the load resistance (R_L) which represents the radio remains constant, at what time will the end point voltage of 0.9 volts be reached? To obtain this answer a graph must be constructed.

To simulate the actual playing of the radio over a period of 8 hours, different values of internal resistance are inserted in the circuit. Immediately after inserting a resistor, take the voltage reading. Remove this resistor and reinsert another resistor and take the voltage reading. Repeat this process for all of the resistors listed. Record all measurements and calculations in the data table.

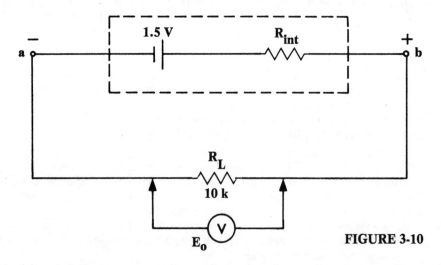

FIGURE 3-10

1. Construct the circuit of Figure 3-10 and connect the voltmeter across R_L, the 10 kΩ load resistor. After each substitution for R_{int}, record the voltage reading E_o across R_L in the data tables. (Disregard the column marked time.)

4.7 Ω	10 Ω	100 Ω	470 Ω	1 kΩ	1.5 kΩ
2.2 kΩ	2.7 kΩ	3.9 kΩ	4.7 kΩ	5.6 kΩ	6.8 kΩ

2. Using 10 × 10 to the inch graph paper, plot the voltage developed across R_L (See Data Table—Step 1—E_o column) versus the elapsed time (See Data Table—Elapsed Time Column). Use the vertical axis for voltage and the horizontal axis for time. From this graph, the end point voltage of the battery can be determined.

26

3. The coordinate points placed on the graph paper should be identified by placing a circle, square, or triangle around them. See Figure 3-11(a). Notice that the curve does not go through all the points, but rather averages them. Figure 3-11(b) is an incorrect method of connecting points. Consult the section on graphing in Appendix C-1 for further assistance.

CORRECT   **INCORRECT**

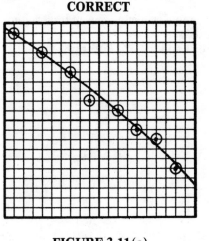   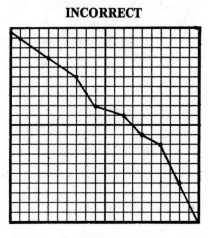

FIGURE 3-11(a)   **FIGURE 3-11(b)**

4. From the graph constructed in Step 2, determine at what time an end point voltage equal to 0.9 volts occurs.
5. Using 4-cycle semi-log paper, construct a graph of the voltage across the load, E_o, (use the linear axis) and the internal resistance, R_{int}, (use the log axis). In the construction of this graph, **use 1 ohm and 0.8 volts as the starting point (origin)** and the values of E_o and R_{int} from the data table of Step 1. Refer to Appendix C-1 for assistance.

SECTION D: VOLTAGE NOTATION

1. Construct the circuit of Figure 3-12. Use three 1.5 volt cells, and 3 different resistor values, 1 kΩ or greater for the three internal resistances.

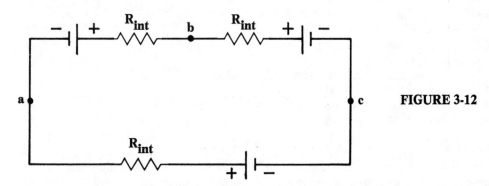

FIGURE 3-12

2. Using the voltage notation, E_{ba}, etc. as described in the introduction, measure the following voltages: E_{ba}, E_{bc}, and E_{ac}. Check the voltage range setting of the voltmeter and remember that the meter is to be placed in parallel with the circuit being tested. (See Figure 3-10 for an example of correct voltmeter placement.)
3. If the meter and circuit were correctly set up, all measurements caused the meter point to deflect up-scale.
4. Using the VOM and the same circuit of Step 1, Figure 3-12, select 3 new values for R_{int} and measure the voltage across the following points: E_{ab}, E_{cb}, and E_{ca}. If the voltage notations are interpreted properly, the meter will deflect down-scale. Since there are no scale divisions in the down-scale direction, the meter leads must be reversed so the meter deflects up-scale. To indicate that the meter originally deflected down-scale, a negative sign is placed in front of the voltage reading.

5. Now that it is understood that voltage may have either a positive or negative value depending upon the reference point, determine both the magnitude and the direction of the voltage across the following points: E_{ca}, E_{bc}, E_{ba}, E_{ab}, E_{ac}, E_{cb}. Use the circuit of Figure 3-12 and three new values for R_{int}.

SECTION E: PHOTOVOLTAIC CELL

The object of this part of the experiment is to construct 2 graphs showing the output voltage from a photovoltaic cell when its distance from the light source is varied.

1. Construct the circuit of Figure 3-13 using a photovoltaic cell and a 1 kΩ load resistor.

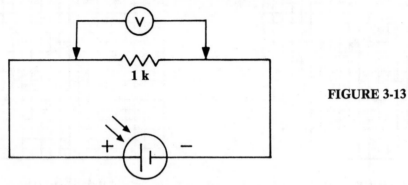

FIGURE 3-13

2. Use a 75 to 100 watt incandescent lamp for a light source. Measure the voltage across the load for distances ranging from 1 to 12 inches from the light source. (Do not leave the cell within 2 inches of the light bulb for over 60 seconds as the cell can be damaged by the heat.) Construct a table of values giving the voltage and corresponding distance. This table will be attached and turned in with the graph.
3. Replace the **load** resistor and repeat Steps 1 and 2 for each of the following resistors: 2.2 kΩ, 4.7 kΩ, and 6.8 kΩ. Make all plots on the same graph paper.
4. Using 10 × 10 to the inch graph paper, graph the measured voltage across the load versus the distance of the cell from the light. Be sure to label your graph properly.
5. Using semi-log paper, repeat Step 4 for the 1 kΩ load resistor.
6. Does shading the cell from the ambient light increase the output voltage?

DATA INTERPRETATION AND CONCLUSIONS

Write a general summary of the ideas presented in this experiment. This discussion should include:
1. The proper procedure for measuring voltage.
2. A description of the resulting circuit conditions when the circuit contains series aiding and series opposing batteries.
3. Voltage conditions for both series and parallel battery hook-up.
4. Your concept of internal resistance.
5. The relationship between voltage and the stimulus of a photovoltaic cell.
6. Your own conclusions.

APPLICATIONS

Batteries are used in (1) radios, (2) television, (3) portable electronic test equipment, (4) citizen's band radio equipment, (5) toys, and many other items. Photovoltaic cells can be used to power (1) roadside emergency phones, (2) rural telephone amplifiers, (3) light-activated toys, (4) satellite power supplies, and (5) transistor radios.

PROBLEMS

1. Using an industrial parts catalog, determine the following:
    a. The cost of any photovoltaic cell.
    b. The specifications called out in the catalog description for this cell.
2. What is the Eveready battery number equivalent to the Neda 1604?
3. Read the manufacturer's descriptive literature in the industrial parts catalog for a battery eliminator. In a short paragraph, describe the battery eliminator. What is the price ? List one application for a battery eliminator.

4 | SERIES CIRCUIT ANALYSIS

OBJECTIVES

Ohm's Law of constant proportionality describes the relationship of voltage, current, and resistance in an electric circuit. It is the basic law for direct current circuit analysis. This experiment stresses Ohm's Law and voltage, current, and resistance laws (derived from Ohm's Law) as the basis for series circuit analysis. Kirchhoff's Voltage Law is also introduced so you will start to think in terms of the voltage loop. A simple technique for determining the internal resistance of a power source is developed. Error and tolerance are considered as a bridge between the theoretical and the practical. The use of the ammeter and current measurements are covered in this experiment.

ELECTRICAL CHARACTERISTICS OF SERIES CIRCUITS

A series circuit is one in which electrical devices and components are connected end-to-end to offer a single path for current to flow from the positive to the negative terminals of an electrical energy source. Figure 4-1 shows three resistors connected in series. The following laws serve as the basis for series circuit analysis:

THE VOLTAGE LAW states: **The sum of the individual voltage drops in a series circuit must equal the voltage that is applied to the series circuit.** Mathematically expressed:

$$E_T = E_1 + E_2 + E_3 + ... + E_n \qquad \text{EQUATION 4-1}$$

where: E_T = the algebraic sum of all batteries or other sources in the series circuit
E_1, E_2, etc. = the voltage drop across R_1, R_2, etc.

THE CURRENT LAW states: **Any current flowing through one component in a series circuit must also flow through the remaining components of the series circuit.** Mathematically expressed:

$$I_T = I_1 = I_2 = I_3 = ... = I_n \qquad \text{EQUATION 4-2}$$

where: I_T = the total current in the series circuit

THE RESISTANCE LAW states: **The total resistance in a series circuit is equal to the sum of the individual resistances.** Mathematically expressed:

$$R_T = R_1 + R_2 + R_3 + ... + R_n \qquad \text{EQUATION 4-3}$$

where R_1, R_2, etc. = the individual resistance values in the series circuit

OHM'S LAW states: **The amount of current flowing in an electrical circuit is equal to the applied voltage divided by the circuit resistance.** Mathematically, this is expressed in the following forms:

$$\text{(a)} \quad I = \frac{E}{R} \qquad\qquad \text{(b)} \quad E = IR \qquad\qquad \text{(c)} \quad R = \frac{E}{I} \qquad \text{EQUATION 4-4}$$

where: E = voltage in volts (V)
I = current in amperes (A)
R = resistance in ohms (Ω)

When any two of the three quantities in Equation 4-4 are known, the third quantity may then be determined.

EXAMPLE 1:

What is the voltage (E) developed across a resistance (R) of 6 ohms when a current (I) of 2 amperes flows through the resistance?

SOLUTION: Use Equation 4-4(b). $E = IR = 2 \text{ A} \times 6\,\Omega = 12$ volts

EXAMPLE 2:

Determine E_2, the voltage drop across resistor R_2, in Figure 4-1.

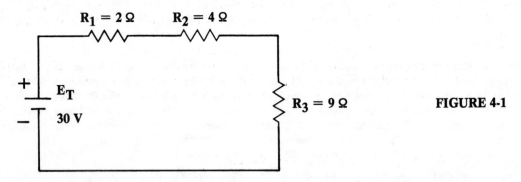

FIGURE 4-1

SOLUTION: The first step in analyzing the problem is to determine what is known about the circuit. Notice that the total voltage applied to the circuit is 30 volts. According to Equation 4-3, the total resistance R_T is 15 Ω. ($R_T = 2\,\Omega + 4\,\Omega + 9\,\Omega = 15\,\Omega$). From Equation 4-4(a), if E_T and R_T are known, then:

$$I_T = E_T/R_T = 30 \text{ V}/15\,\Omega = 2 \text{ A}$$

From Equation 4-4(b), E_2 can be determined if I_2 and R_2 are known, Equation 4-2 states that $I_T = I_2$. Therefore,

$$E_2 = I_T R_2 = 2 \text{ A} \times 4\,\Omega = 8 \text{ V}$$

Another equation that is of great assistance in determining the voltage across any resistance in a series circuit is the **voltage divider equation.**

$$E_X = \frac{E_A R_X}{\Sigma R}$$

EQUATION 4-5

where: E_A = voltage applied to the series circuit

E_X = voltage developed across the selected resistor

R_X = the selected resistor

ΣR = sum of all the resistors in the series circuit

Σ = (sigma) Greek letter used to indicate summation

EXAMPLE 3:

Using Equation 4-5, determine the voltage across R_3 in Figure 4-1.

SOLUTION: The formula is written as follows for the required solution. Notice that the subscript "X" was replaced by a 3.

$$E_3 = \frac{E_A R_3}{\Sigma R} = \frac{30 \text{ V} \times 9\,\Omega}{15\,\Omega} = 18 \text{ V}$$

Equation 4-5 is used whenever possible because it reduces the mathematical burden in circuit analysis.

KIRCHHOFF'S VOLTAGE LAW AND THE VOLTAGE LOOP

KIRCHHOFF'S VOLTAGE LAW states: **The algebraic sum of all the voltages around a closed loop must equal zero.**

Using this law and the voltage notation concept developed in Experiment 3, analyze the circuit of Figure 4-2. Assume that the circuit has been set up on a test bench and that a voltmeter is being used to make the measurements. The circuit analysis should go something like this:

1. Determine whether the direction of movement will be clockwise (CW) or counterclockwise (CCW) around the circuit. Once a direction has been selected and the analysis started, the direction cannot be reversed.
2. Determine from which point the analysis is to start—i.e. **a**, **b**, **c**, **d**, etc.
3. Place the negative lead on that point. The placement of the positive lead depends upon the selected direction. For the circuit of Figure 4-2, Point **a** will be assumed as the starting point and the direction of analysis will be in the clockwise direction. Therefore, the positive lead will be placed at Point **b**.

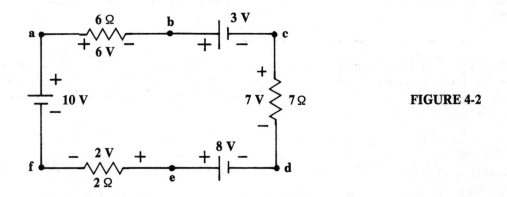

FIGURE 4-2

4. Record the measured voltage and polarity; then, place the negative lead on Point **b** and the positive lead on Point **c**. Record the voltage and polarity, etc.

Since the analysis was started at Point **a** and movement was in the clockwise direction, the readings obtained were:

$$E_{ba} = -6\ V, \qquad E_{cb} = -3\ V, \qquad E_{dc} = -7\ V,$$

$$E_{ed} = +8\ V, \qquad E_{fe} = -2\ V, \qquad E_{af} = +10\ V.$$

If Kirchhoff's law is valid, then the summation of these voltages must equal zero. Therefore, it can be stated that:

$$E_{ba} + E_{cb} + E_{dc} + E_{ed} + E_{fe} + E_{af} = 0$$

Substituting the voltages:

$$(-6\ V) + (-3\ V) + (-7\ V) + (8\ V) + (-2\ V) + (10\ V) = 0$$

Had the counterclockwise direction been selected, the notation would have been:

$$E_{fa}, E_{ef}, E_{de}, \text{etc.}$$

Figures 4-3(a) and (b) show how the circuit components can be regrouped and combined to provide an equivalent circuit. Note that the net current is 1 A as shown in Figure 4-3(b).

31

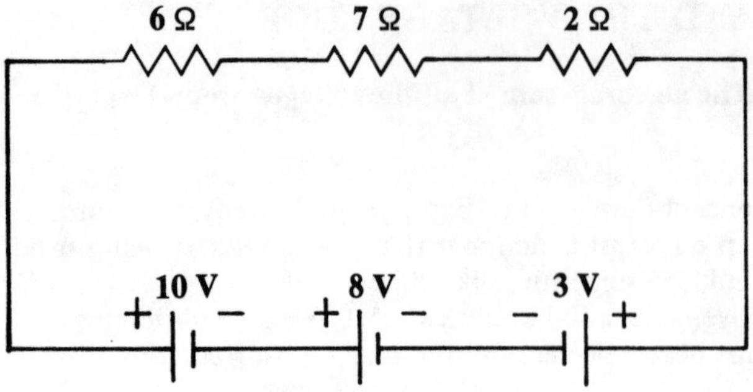

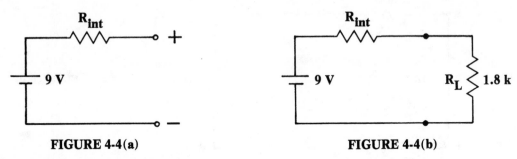

FIGURE 4-3(a) 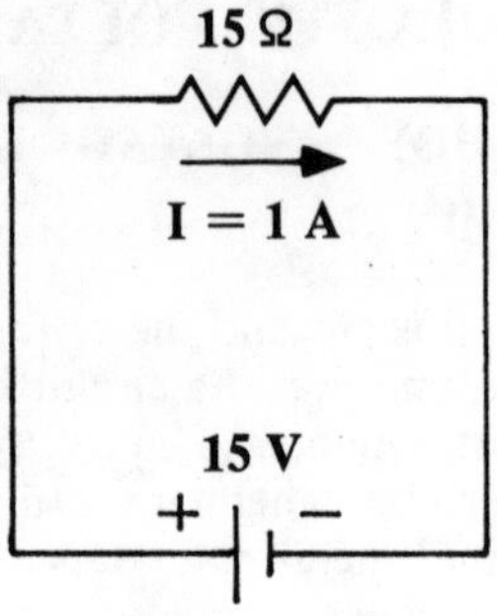   FIGURE 4-3(b)

DETERMINING THE INTERNAL RESISTANCE OF A POWER SOURCE

In Experiment 3, it was stated that a technique would be developed to determine the value of the internal resistance of a voltage source. The circuits of Figure 4-4(a) and 4-4(b) are used to develop this concept.

FIGURE 4-4(a)   FIGURE 4-4(b)

NOTE: Although this technique is satisfactory for many power sources, a more sophisticated test is required for the well-regulated power supply.

Assume that the battery in Figure 4-4(a) is powering a transistor radio that looks like the 1.8 kΩ load of Figure 4-4(b) and that the radio will quit playing if the load voltage is reduced to 5 volts. Using a DC volt meter, the first measurement of the open circuit battery voltage (Figure 4-4(a)) was measured as 9 V. Assume that after placing the 1.8 kΩ load on the battery, Figure 4-4(b), the voltage across the load was measured as 5 volts. Using Kirchhoff's voltage law and Figure 4-4(b), it can be seen that if the open circuit battery voltage is 9 volts and the voltage developed across the 1.8 kΩ load is 5 volts, then 4 volts must be dropped across the internal resistance (R_{int}). Notice how Kirchhoff's voltage law, in conjunction with the closed-loop concept, is used to aid in the analysis of this circuit. In order to determine the internal resistance, Equation 4-6 will be used.

$$R_{int} = \left[\frac{E_s}{E_o} - 1\right] R_L \qquad \text{EQUATION 4-6}$$

where: R_{int} = internal resistance
$\quad\quad\;\; R_L$ = load resistance
$\quad\quad\;\; E_s$ = source voltage
$\quad\quad\;\; E_o$ = voltage across load

Substituting 9 V for E_s, 5 V for E_o, and 1.8 kΩ for R_L, the result is:

$$R_{int} = \left[\frac{9\text{ V}}{5\text{ V}} - 1\right] 1.8\text{ k}\Omega = 1.44\text{ k}\Omega$$

32

ERROR AND TOLERANCE

Invariably, a difference will occur between the theoretical computations and the actual measurements. This difference is defined as error and is expressed in terms of "percentage error".

$$\% \text{ error} = \frac{MV - CV}{CV} \times 100$$

EQUATION 4-7

where: CV = computed value (theory)
MV = measured value (practical)

EXAMPLE 4:

The computed current of a certain circuit was 1.8 mA. The meter reading of the milliammeter was 1.7 mA. Is this reading "close enough"?

SOLUTION:

$$\% \text{ error} = \frac{MV - CV}{CV} \times 100 = \frac{1.7 \text{ mA} - 1.8 \text{ mA}}{1.8 \text{ mA}} \times 100 = -5.6 \%$$

Normally errors of 10% or less are acceptable.

THE VARIABLE POWER SUPPLY

Included in this experiment is a new symbol, see Figure 4-5, which is a Variable Power Supply (VPS). This type of power supply will be used in this and subsequent experiments. The exact nature of this supply may not always be the same, but it will always indicate that whatever type of supply is used it must be variable.

FIGURE 4-5

NOTE: In **industry** it is common practice to call a voltage source a power supply.

SUGGESTED OBSERVATIONS
1. Think the problem through prior to taking action.
2. Make certain that the ammeter is properly connected in the circuit prior to energizing the circuit.
3. Always check the meter range setting prior to using the meter.
4. **Never** make circuit changes with the circuit energized.
5. Make the bench set-up, circuit construction, and notes as neat as possible.
6. Record all readings and calculations in the data table.

LIST OF MATERIALS   Alternate materials
1. VOM with manual
2. Resistors—at least ½ watt (value in ohms)

120 Ω	150 Ω	180 Ω	220 Ω	330 Ω
470 Ω	1 kΩ	1.2 kΩ	1.5 kΩ	1.8 kΩ
2.2 kΩ	2.7 kΩ	3.3 kΩ	3.9 kΩ	4.7 kΩ

3. Variable Power Supply (VPS)
4. Three 1.5 volt batteries
5. Graph Paper (10 x 10 or 20 x 20 to the inch)

EXPERIMENTAL PROCEDURE:

In this section, current measuring techniques will be developed for the VOM. Current measurement is not difficult, but if done improperly, current measurements may result in a damaged meter. This factor can be directly related to a lack of understanding and inattentiveness on the part of the student. Numerous damaged meters in every introductory direct current circuit laboratory bear testimony to this fact.

Suppose, for example, it is necessary to determine the current flowing through R_2 of Figure 4-1. The first questions that should be asked are: "Where is the circuit broken (opened) to insert the ammeter for this measurement?" and "Will this meter handle the circuit current?" The student must answer these questions before he can proceed and take the measurement. It must be remembered that (1) the ammeter must be connected in **SERIES** with the other components, and (2) the current range of the meter must be set up prior to applying the power to the circuit. If the approximate amount of current flowing in a circuit is unknown, **DO NOT INSERT AN AMMETER TO FIND OUT!** This statement cannot be overemphasized.

Consider the following situation: the starter motor on an automobile is not functioning properly. The decision is made, by an **inexperienced person**, to measure the current. The starter cable is disconnected and a VOM is inserted into the circuit. Although the exact current the starter should take is unknown, it is decided that the highest meter range, 10 or 12 amperes, will be sufficient for this measurement. When the starter motor was energized, the meter was completely destroyed because the starter required 150 amperes for proper operation. Obviously, had the starter current requirement been considered realistically, this measurement would not have been attempted using a meter with such a limited current range.

Many other examples can be cited to show what happens because of a lack of planning. Referring back to the circuits of Figure 4-1 and 4-2, it was possible to estimate the circuit current. Consequently, a meter capable of handling at least 2 amperes would be required for Figure 4-1 and a meter with capabilities of 1 ampere or better for the circuit of Figure 4-2. Remember—always **THINK** before taking action!

SECTION A: CURRENT MEASUREMENT

1. Set the function selector to DC mA and set the range switch to 10 mA.
2. Insert the test leads in the proper jacks for measuring DC current. The black lead is the common lead and it is placed in the jack labeled common (negative). Consult the operator's manual or check with the instructor for a detailed set-up procedure.
3. Construct the circuit of Figure 4-6(a).

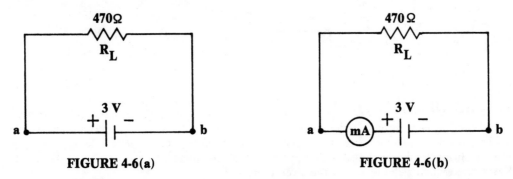

FIGURE 4-6(a)   FIGURE 4-6(b)

4. Measure the **current** flowing through Point **a** by **breaking** the circuit at Point **a**, see Figure 4-6(b), and **inserting** the milliammeter in **series** with the load and voltage source. Connect the common lead to the 470 Ω resistor, and touch the positive probe to the positive battery terminal. If the meter pointer is deflected beyond the full scale value, **remove** the probe and increase the range setting. Record the measurement.
5. Repeat Steps 1 through 4 for a current measurement at Point **b**.

SECTION B: OHM'S LAW

1. Construct the circuit of Figure 4-7. Use a 1.2 kΩ and a 1.8 kΩ resistor for R_1 and R_2 respectively. Using a 0 - 10 milliammeter, insert it in the circuit at Point **a**. Observe meter polarity.

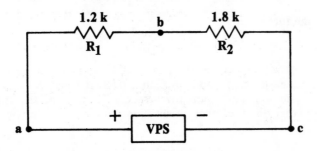

FIGURE 4-7

2. With the VPS set to zero, increase the supply voltage for a milliammeter reading of 6 mA. Without disturbing this adjustment, remove the power from the circuit, and **THEN** remove the milliammeter from the circuit. Be certain to reconnect Point **a**.

3. Reapply the power to the circuit and measure the voltage across Points **a** and **c** (E_{ac}).

4. Remove the circuit from the VPS and measure the resistance across Points **a** and **c**.

5. Using Ohm's law and the value of resistance and current determined in Steps 2 and 4, calculate the supply voltage. $E_{ac} = IR_T$.

6. Compute the % difference between the measured and calculated values of source voltage. Use Equation 4-7 and the values determined in Steps 3 and 5.

SECTION C: GRAPHIC SOLUTION OF OHM'S LAW PROBLEMS

In this portion of the experiment, several values of resistance will be used to make a graph showing current in relation to voltage. From any selected curve, the student will be able to determine either or both current and voltage.

1. Construct the circuit of Figure 4-8. Use a load resistance (R_L) of 1 kΩ.

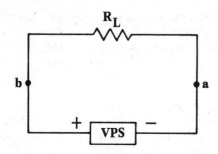

FIGURE 4-8

2. Adjust the VPS voltage, E_{ba}, for a 1 volt reading on the VOM. Break the circuit at Point **a** and use the ammeter section of the meter to measure the current.

3. Repeat Step 2 for VPS voltages of 2, 3, 4,...,10 volts.

4. Replace the 1 kΩ load resistor with a 1.2 kΩ resistor, and repeat Steps 2 and 3.

5. Replace the 1.2 kΩ load resistor with a 1.5 kΩ resistor, and repeat Steps 2 and 3.

6. Plot a curve of current (vertical axis) versus voltage (horizontal axis) for each of the load resistors used in the circuit of Figure 4-8. Use 10 × 10 to the inch graph paper. Consult Appendix C-1, "Graphing Techniques". Include this graph in the experimental write-up.

7. There should now be three linear curves which represent the value of each load resistance used. Label each curve with its values of load resistance (1 kΩ, 1.2 kΩ, 1.5 kΩ). These curves can be used to determine the current when the voltage is known and vice versa.

8. Using the graph of Step 6, determine the values of the following unknown quantities:
   Find: a) E when R = 1 kΩ   and I = 5.25 mA
          b) E when R = 1.5 kΩ and I = 2.85 mA
          c) I when E = 8.2 V and R = 1.2 kΩ
          d) I when E = 3.85 V and R = 1.5 kΩ

SECTION D: SERIES EQUIVALENT RESISTANCE

A complex series circuit can be reduced to a simple circuit consisting of one voltage source and one resistance. The equivalent circuit is equal to the original circuit if the equivalent circuit parameters (current, resistance, and voltage) produce the same results as the original circuit parameters. In this section the circuit parameters of Figure 4-9(a) will first be determined; then the original circuit will be replaced with an equivalent circuit, Figure 4-9(b); and finally, the difference between the parameters of the two circuits will be determined.

1. Before connecting the VPS to the circuit of Figure 4-9(a), measure the total resistance between Points **a** and **b**.

2. Insert the VOM with the range selector set to the 10 mA range and connect the VPS in the circuit of Figure 4-9(a).

3. With the VPS turned completely down, turn the supply on. Slowly adjust the VPS until the milliammeter reads 8 mA. Do not move the voltage adjustment once 8 mA has been obtained.

4. Turn off the VPS and remove the milliammeter from the circuit. Reconnect the circuit (without the milliammeter) and measure the source voltage (E_{ba}).

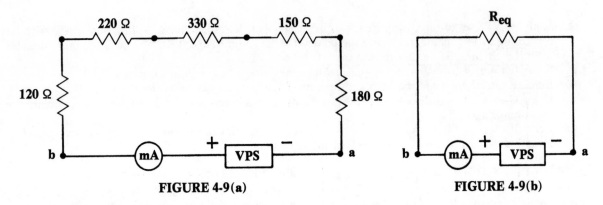

FIGURE 4-9(a)   FIGURE 4-9(b)

5. Replace the five resistors of Figure 4-9(a) with one resistor which has a value equal to the measured resistance of Step 1. This resistance is equal to the total resistance, R_T, and it is now called the equivalent resistance, R_{eq}.
6. Connect the circuit of Figure 4-9(b) using the equivalent resistance from Step 5.
7. Repeat Steps 1 through 4 only this time use Figure 4-9(b).
8. Calculate the % difference between: (1) values of the source voltage, (2) current, and (3) total resistance of the two circuits.
9. Based upon the calculation of Step 8, are the two circuits equivalent?

SECTION E: KIRCHHOFF'S VOLTAGE LAW

The next portion of this experiment uses symbolic notation to show that the algebraic sum of the voltage drops around a closed loop is zero.

1. Construct the circuit of Figure 4-10. After the circuit is connected, set the VPS to 20 volts.

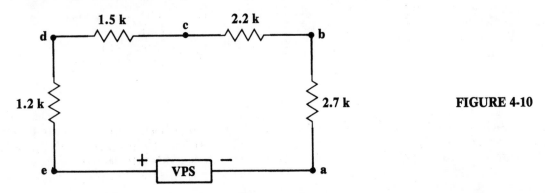

FIGURE 4-10

2. Starting at Point **a** and moving in a counterclockwise direction around the closed loop, measure the voltage drops of E_{ba}, E_{cb}, E_{dc}, E_{ed}, and E_{ae}. Remember that the second subscript designates the placement of the common meter lead. Polarity must be recorded with the voltage readings.
3. Complete the following equation using the value and direction of voltages determined in Step 2. Replace the voltage symbol with the magnitude and place either a plus sign or a minus sign to indicate direction between the voltage symbols. Record these answers in the data table.

$$E_{ba} (\pm) E_{cb} (\pm) E_{dc} (\pm) E_{ed} (\pm) E_{ae} = 0$$

4. Replace the resistors of Figure 4-10 with the following resistors: 2.7 kΩ, 3.3 kΩ, 3.9 kΩ, 4.7 kΩ. Repeat Steps 1 through 3 moving around the loop in a clockwise direction.

SECTION F: A MULTIPLE SOURCE VOLTAGE LOOP

1. Construct the circuit of Figure 4-11. Use dry cells and a VPS for the voltage sources.
2. Measure the voltage drops around the loop starting with Point **b** and progressing in a counterclockwise direction. (E_{cb}, E_{dc}, etc.). Record both the magnitude and the polarity of the voltage drops.
3. Using Kirchhoff's voltage law, show that the algebraic sum of the voltages around the circuit of Figure 4-11 is zero.

36

4. Repeat Steps 1 through 3, only this time make the measurements in a clockwise direction. (E_{ab}, E_{ha}, E_{gh}, etc.).

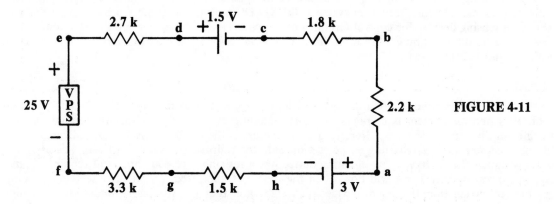

FIGURE 4-11

SECTION G: DETERMINING INTERNAL RESISTANCE

Equation 4-6, given earlier in the introductory discussion, along with Figure 4-12 form the basis of this experiment. Equation 4-6 stated:

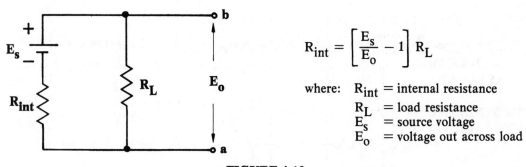

$$R_{int} = \left[\frac{E_s}{E_o} - 1 \right] R_L$$

where: R_{int} = internal resistance
R_L = load resistance
E_s = source voltage
E_o = voltage out across load

FIGURE 4-12

1. Relating Equation 4-6 to Figure 4-12, the source voltage (E_s) can be determined if the load (R_L) is removed and the voltage is measured across the Terminals **a** and **b**. R_L is picked by the student and the output voltage (E_o) is measured across the Terminals **a** and **b** with the load resistor connected. Since all variables of Equation 4-6 are known, the internal resistance (R_{int}) can then be calculated.
2. The internal resistance of a voltage source will now be determined. Construct the circuit of Figure 4-12 using dry cells for the source of 3 volts. A load resistor is picked arbitrarily. Since a dry cell has an extremely small internal resistance, R_{int} will be simulated by placing a 1 kΩ resistor in series with the battery. A reasonable value for the load resistor would be 1.5 kΩ.
3. Using Step 1 and Equation 4-6, experimentally determine the value of R_{int}.
4. Compare the internal resistance value determined in Step 3 to that given in Step 2. Use Equation 4-7 to determine the % error.
5. Replace the 1 kΩ internal resistance with an internal resistance of 2.7 kΩ, and then repeat Steps 1 through 4. This time you will pick the value for load resistance.

DATA INTERPRETATION AND CONCLUSIONS

Write a general summary of the ideas presented in this experiment on Series Circuit Analysis. This discussion should include:
1. The relationship of Kirchhoff's voltage law to the series circuit.
2. The fundamental laws for the series circuit.
3. The methods used to determine the value of current.
4. The need for a method of comparing theoretical and experimental values.
5. Your own conclusions.

APPLICATIONS

The series circuit has numerous applications throughout the electronic and electrical fields. These include: (1) street light connections, (2) telephone control circuitry, (3) computer circuits, (4) some burglar alarm systems, (5) some fire alarm systems, (6) AD-DC motors, (7) some types of Christmas tree lights, (8) some sensing devices for control circuitry on machinery, (9) the meter multiplier circuits, (10) voltage reduction circuits—running a 6 V radio from a 12 V battery, (11) one of the methods of connecting loud speakers, and (12) light switches.

PROBLEM

A technician has been employed by a lumber and logging company to maintain their communication equipment. His first assignment is to repair a faulty signaling circuit between the lumbermill and the logging camp—a distance of 25 miles. Referring to the notes on the original installation indicates that a pair of exposed copper wires were strung on poles alongside the railroad track. No. 14 wire was used and has a total resistance of 700 ohms. The signaling device has a resistance of 800 ohms and requires 10 mA to operate. At present, only 1 mA will flow when the device is energized. A trip to the logging camp convinces the technician that the signal device is all right and that the trouble is undoubtedly a high resistance contact in the wire. He makes a rough sketch of the entire layout between the camp and mill area as follows:

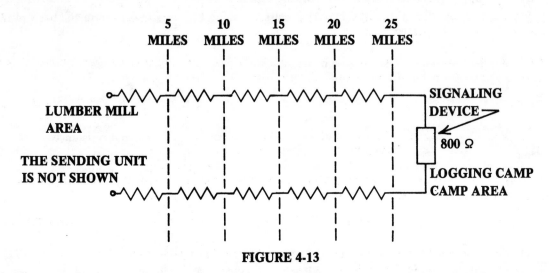

FIGURE 4-13

Using his sketch, explain the procedure you (as the technician) would follow to determine where the high resistance contact is located. Remember, it is very unlikely that a poor connection would be spotted while riding along looking at 25 miles of wire.

5 || PARALLEL CIRCUIT ANALYSIS

OBJECTIVES

Three laws involving voltage, current, and conductance serve as the basis for parallel circuit analysis. This experiment stresses the application of those laws and the use of conductance. Kirchhoff's Current Law is introduced along with the current arrow.

ELECTRICAL CHARACTERISTICS OF PARALLEL CIRCUITS

A parallel circuit is one in which electrical devices are connected to form two or more current paths. Figure 5-1 shows two resistors connected in parallel. In a way similar to the description of the series circuit, it is also possible to describe and define a parallel circuit. There are three basic laws that describe the parallel circuit.

THE VOLTAGE LAW states: **The voltage across any branch of a parallel circuit is equal to the voltage applied to the parallel circuit.** Mathematically expressed:

$$E_T = E_1 = E_2 = E_3 = ... = E_n \qquad \text{EQUATION 5-1}$$

where: E_T = the total voltage applied to the parallel circuit
E_1, E_2, etc. = the voltage across each branch of the parallel circuit

THE CURRENT LAW states: **The total current entering or leaving a parallel circuit is equal to the sum of the currents of the individual branch circuits.** Mathematically expressed:

$$I_T = I_1 + I_2 + I_3 + ... + I_n \qquad \text{EQUATION 5-2}$$

where: I_T = the total current entering or leaving the parallel circuit
I_1 = the current within branch 1
I_2 = the current within branch 2, etc.

THE CONDUCTANCE LAW states: **The total conductance of the parallel circuit is equal to the sum of the individual branch conductances.** Mathematically expressed:

$$G_T = G_1 + G_2 + G_3 + ... + G_n \qquad \text{EQUATION 5-3}$$

where: G_T = the total conductance of the parallel circuit
G_1 = the conductance of branch 1
G_2 = the conductance of branch 2, etc.

NOTE: The unit of conductance is the **siemens (S)**.

Conductance is defined as: "the ease with which a material will allow current to flow". It is mathematically related to resistance by the following equations:

$$G_T = \frac{1}{R_T}, \qquad\qquad G_1 = \frac{1}{R_1}, \qquad\qquad G_2 = \frac{1}{R_2}, \text{ etc.} \qquad \textbf{EQUATION 5-4(a)}$$

where: G_T = the total conductance
R_T = the total resistance
G_1 = the conductance of branch 1
R_1 = the resistance of branch 1

It is obvious from Equation 5-4(a) that conductance is the reciprocal of resistance and vice versa. Hence:

$$R_T = \frac{1}{G_T}, \qquad\qquad R_1 = \frac{1}{G_1}, \text{ etc.} \qquad\qquad \textbf{EQUATION 5-4(b)}$$

Substituting Equation 5-4(b) into Equation 4-4(b), $E = IR$, results in:

$$E_T = \frac{I_T}{G_T}, \qquad\qquad E_1 = \frac{I_1}{G_1}, \text{ etc. and} \qquad \textbf{EQUATION 5-5(a)}$$

$$I_T = E_T G_T, \qquad\qquad I_1 = E_1 G_1, \quad \text{etc.} \qquad\qquad \textbf{EQUATION 5-5(b)}$$

The following example uses these equations to solve for circuit parameters.

EXAMPLE 1:

Figure 5-1 shows a simple parallel circuit. Determine I_T, I_1, I_2, R_T, and G_T.

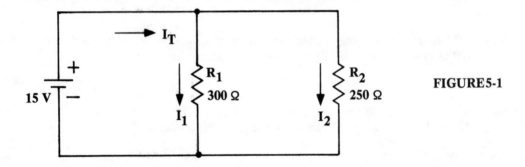

FIGURE 5-1

SOLUTION: Since the voltage across R_1 and R_2 is the same (Equation 5-1), then by Equation 4-4(a):

$$I_1 = \frac{E_1}{R_1} \qquad\qquad \text{and} \qquad\qquad I_2 = \frac{E_2}{R_2}$$

$$I_1 = 15\,V/300\,\Omega = 50\,mA \qquad \text{and} \qquad I_2 = 15\,V/250\,\Omega = 60\,mA$$

By Equation 5-2, $\qquad I_T = I_1 + I_2$; therefore,

$$I_T = 50\,mA + 60\,mA = 110\,mA$$

By Equation 4-4(c): $\qquad R_T = E_T/I_T = 15\,V/110\,mA = 136.4\,\Omega$

By Equation 5-4(a), $\qquad G_T = 1/R_T = 1/136.4\,\Omega = 7.33\,mS$

An alternate method for obtaining G_T is by Equation 5-3 and Equation 5-4(a).

$$G_T = G_1 + G_2, \qquad\qquad G_1 = \frac{1}{R_1}.$$

Therefore, by substitution:

$$G_T = \frac{1}{R_1} + \frac{1}{R_2} = \frac{1}{300\,\Omega} + \frac{1}{250\,\Omega} = 7.33\,mS$$

When only two resistors are in parallel, Equation 5-6 can be used to solve for R_T.

$$R_T = \frac{R_1 R_2}{R_1 + R_2} \qquad\qquad\qquad \textbf{EQUATION 5-6}$$

Equation 5-7 can be used to solve for R_T when any number of resistors are in parallel.

$$R_T = \cfrac{1}{\cfrac{1}{R_1} + \cfrac{1}{R_2} + \cfrac{1}{R_3}}$$

EQUATION 5-7

KIRCHHOFF'S CURRENT LAW AND THE CURRENT ARROW

The preceding examples describe parallel circuit parameters. However, to gain a complete understanding of parallel circuit analysis, the concepts of Kirchhoff's Current Law and the current arrow must also be introduced.

KIRCHHOFF'S CURRENT LAW states: **The algebraic sum of the current entering a node (junction) is zero.** Stated another way, the current entering a node must be equal to the current leaving the node.

NOTE: A **minor node** is defined as the connecting point of two components, e.g., the connection of the battery and resistor shown in Figure 5-2. A **major node** is defined as having three or more components connected together. See Figure 5-3. Since minor nodes are of little significance, any further reference to nodes will imply that they are major nodes.

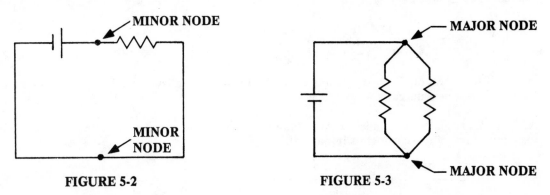

FIGURE 5-2   **FIGURE 5-3**

The concept of indicating current flow with the current arrow is directly related to Kirchhoff's current law as shown in Figure 5-4.

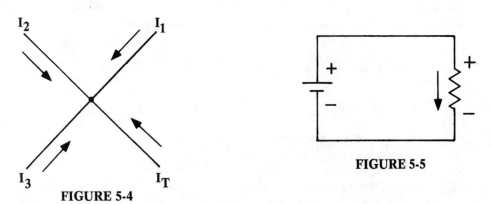

FIGURE 5-4   **FIGURE 5-5**

Since conventional current flows from positive to negative (external to the source), it can be stated that the head of the arrow is negative and the tail is positive. See Figure 5-5. The significance of this concept will become more apparent in each subsequent experiment. The current arrow is shown in the following example:

EXAMPLE 2:

Given the circuit of Figure 5-6, with an I_T of 198.4 mA, determine the value of I_1.

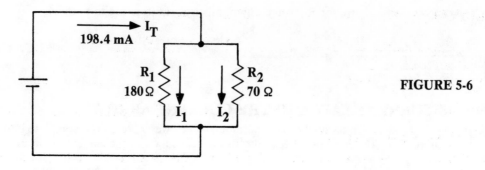

FIGURE 5-6

SOLUTION: The equivalent parallel conductance (19.84 mS) of resistors R_1 and R_2 is determined by Equation 5-3. Dividing the total current (198.4 mA) by the total conductance results in:

$$E_T = \frac{I_T}{G_T} = \frac{198.4 \text{ mA}}{19.84 \text{ mS}} = 10 \text{ V}$$

Therefore, by Equation 4-5(a):

$$I_1 = \frac{E_1}{R_1} = \frac{10 \text{ V}}{180 \text{ }\Omega} = 55.6 \text{ mA}$$

ALTERNATE SOLUTION: By using the **current divider equation**, the labor involved in the solution to Example 2 can be reduced.

NOTE: Equation 5-8(a) is valid for a maximum of two resistors in parallel while Equation 5-8(b) is valid for a maximum of two conductors in parallel.

$$I_X = \frac{I_T R}{\Sigma R} \qquad \textbf{EQUATION 5-8(a)} \qquad\qquad I_X = \frac{I_T G}{\Sigma G} \qquad \textbf{EQUATION 5-8(b)}$$

where: I_T = total current flowing into the parallel circuit
I_X = current to be determined
R = resistance that I_X is not flowing through
ΣR = sum of the two resistances
G = conductance that I_X is flowing through
ΣG = sum of the conductances

Substituting the values of Example 2 into Equation 5-8(a) results in:

$$I_1 = \frac{I_T R_2}{R_1 + R_2} = \frac{198.4 \text{ mA} \times 70 \text{ }\Omega}{180 \text{ }\Omega + 70 \text{ }\Omega} = 55.6 \text{ mA}$$

SUGGESTED OBSERVATIONS

1. Before placing a meter into the circuit, first approximate the circuit parameters. Then, set the range switch to an appropriate level.
2. Remember that the **voltmeter** is connected across the load and the **milliammeter** is connected in series with the load.
3. Do not operate the lamps above their rated voltage.
4. Record all measurements and calculations in the data tables.

LIST OF MATERIALS

1. VOM
2. 3 No. 47 Lamps with Holders
3. 4 1.5 volt Batteries
4. Resistors—all at least ½ watt
   3.3 kΩ   4.7 kΩ   6.8 kΩ
5. Variable Power Supply (VPS)

Alternate Materials

3 No. 40 Lamps with Holders

_____

_____

_____

EXPERIMENTAL PROCEDURE

This experiment explores the concepts of the parallel circuit including Kirchhoff's Current Law, current division and current notation.

SECTION A: PARALLEL CIRCUIT CHARACTERISTICS

1. Construct a series circuit, as in Figure 5-7, consisting of four 1.5 volt dry cells (connected in series to provide 6 V and one No. 47 pilot lamp.

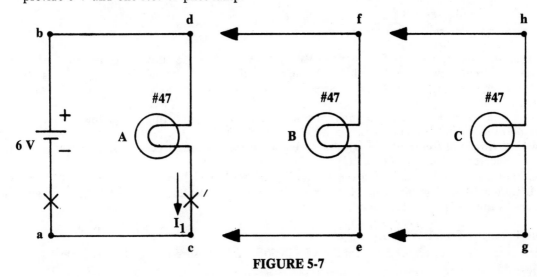

FIGURE 5-7

2. Measure the total current by breaking the circuit at Point **a** (**break Point "X"**) and inserting a 0 -1000 milliammeter.
3. Add an additional lamp (Lamp B) in parallel with the circuit across nodes **c** and **d**. Record the meter reading.
4. Connect lamp C in parallel with the circuit across nodes **e** and **f**. Record the meter reading.
5. Why did the total circuit current increase when additional loads were placed across the voltage source?
6. Remove the meter from the circuit at Point **a** and reconnect the circuit. Open the circuit at the break point "X'" between Nodes **c** and **d** in Figure 5-7 and insert a 0 -1000 milliammeter. With all three lamps connected in parallel, observe the meter reading as Lamp C is removed and then Lamp B is removed. Comment on the brightness of Lamp A. Did the current and brightness of Lamp A change when Lamps B and C were removed?
7. Reconnect the three lamps in parallel with the source voltage. Measure the following voltage drops:

E_{ba}, E_{dc}, E_{fe}, E_{hg}

8. In the circuit of Step 7, what value of voltage would the meter indicate if the measurement was taken across Points h to a? How would this compare with the value read from Points b to g?
9. Place a voltmeter across Lamp A and observe the voltage reading as Lamps C and B are removed from the circuit. Did the voltage change?
10. In a short statement, account for the result in Step 9.
11. If the lamps were connected in series, would the voltage across one of the lamps remain constant when additional lamps are placed in series in the circuit? Why? If you are unable to answer this question, then set up a series circuit and determine the results.

43

SECTION B: KIRCHHOFF'S CURRENT LAW

1. Using the resistance values given in Figure 5-8 and the source voltage of 10 V, calculate the total current (I_T) and each of the branch currents (I_1, I_2, I_3).

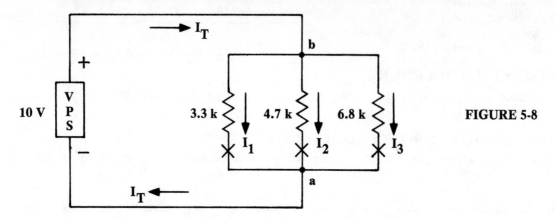

FIGURE 5-8

2. Construct the circuit of Figure 5-8. Set the VPS to 10 volts after the circuit is constructed. Measure the total current flowing in the circuit by opening the circuit below Point **a** and inserting a milliammeter. Be sure to set the range selector to a value high enough to measure the calculated current.)

3. Measure and record each of the branch currents (I_1, I_2, I_3) by inserting the milliammeter at each break point indicated in Figure 5-8.

4. Show that Kirchhoff's current law is valid by equating the measured current(s) into Node **a** to the measured current(s) leaving Node **a**. ($I_T = I_1 + I_2 + I_3$)

5. Repeat Steps 2 through 4 for the currents entering and leaving Node **b**.

6. Use Equation 4-7 to determine the % difference between the calculated value of I_T (Step 1) and the measured value (Step 2) of I_T.

SECTION C: CURRENT DIVISION

1. Construct the circuit of Figure 5-9. Set the VPS to 30 V and measure the total current flowing in this circuit.

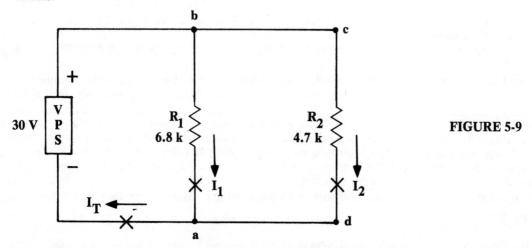

FIGURE 5-9

2. Using the current of Step 1 and Equation 5-8, determine the current, I_1, through the 6.8 kΩ resistance.

3. Open the circuit between Points **a** and **b** and measure current, I_1.

4. Repeat Steps 2 and 3 for the branch containing the 4.7 kΩ resistor.

5. Add the calculated branch currents and compare this sum to the sum obtained when the measured branch currents were added. Use Equation 4-7.

44

SECTION D: PARALLEL CIRCUIT VOLTAGE LOOPS

In Experiment 4, it was found that the algebraic sum of the voltages around a closed loop was equal to zero. Since a series circuit is actually one closed voltage loop, there was no concern with what constituted a voltage loop. However, in the parallel circuit, there are always two or more loops. The parallel circuit of Figure 5-10 has 3 loops as noted.

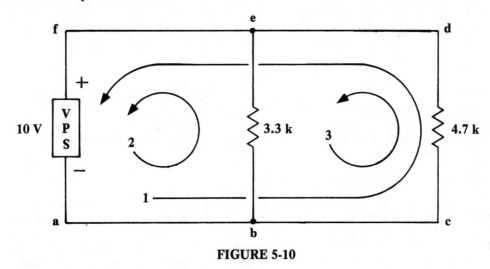

FIGURE 5-10

1. Construct the circuit of Figure 5-10.
2. Using the VOM measure the voltage around Loop 1. (E_{ba}, E_{cb}, E_{dc}, E_{ed}, E_{fe}, E_{af})
   When recording the answers, remember to write down both the magnitude and polarity of the voltage.
3. Using Kirchhoff's voltage law, show that the algebraic sum of the voltages in Step 2 are equal to zero.
4. Repeat Steps 2 and 3 for voltage Loop 3.

DATA INTERPRETATION AND CONCLUSIONS

Write a general summary of the ideas presented in this experiment. This discussion should include:
1. A comparison of series and parallel circuits, as to total current and total voltage.
2. The advantage of using a parallel circuit for lighting.
3. An accounting of the difference between the calculated and measured values of current in Sections B and C.
4. Your own conclusions.

APPLICATIONS

Parallel circuitry is used exclusively in house and automotive wiring. In addition, parallel circuits are used extensively in such devices as radios and televisions.

PROBLEMS

1. Using an industrial parts catalog, determine the voltage and current specification for a N_0. 47 lamp. In addition, list two other lamps and their specifications.
2. Explain why it is desirable to use parallel circuits in house wiring.
3. What conclusion can be arrived at when comparing the values of the total resistance of a parallel circuit to the values of the individual branch resistances?
4. With respect to conductance, how do the branch currents of a parallel circuit divide?
5. Outline the procedure used to determine where a short circuit might exist in a parallel string of lights. This string consists of 8 lamps and is used to illuminate a picnic area in a park. All wiring and lamps are easily accessible. **Consider all possibilities.** It is recommended that a circuit sketch be made to assist in evaluating this problem.

6 | COMPOUND CIRCUIT ANALYSIS

OBJECTIVES

Electric circuits that are formed by combining series and parallel circuits are called **compound circuits**. This experiment explains how compound circuits are analyzed by sectioning into series and parallel circuits. Continued application is made of both current arrows and voltage loops with emphasis placed on schematic drawing and troubleshooting.

ANALYSIS OF COMPOUND CIRCUITS

Since the compound circuit is a combination of series and parallel circuits (see Figure 6-1), its analysis may be simple or extremely complex. The degree of complexity of any circuit is usually determined by the amount of time spent, and the level of mathematics required for its resolution.

For analysis, a compound circuit must first be sectioned into individual series and parallel circuits. Each series section is then treated as a series circuit and all series circuit laws can be utilized to obtain the solution. Those sections containing parallel circuits are solved using the parallel circuit laws.

The following steps outline the general procedure required to section a compound circuit.
1. Determine what information is given.
2. Determine what is required for the solution.
3. Draw a schematic that represents the given conditions and label each known component with its value.
4. Draw in current arrows and observe where the currents divide (major nodes).
5. Noting current divisions, section the circuit into series and parallel groups. Number each section.

NOTE: When current divides and flows through several paths, it must eventually recombine at another point. Therefore, a parallel section exists between the point of separation and the point of recombination. It is entirely possible to have series and series-parallel circuits within parallel sections.

6. Determine any additional information that can be obtained from the given data that might be of assistance in the solution of the problem.
7. Obtain the solution.

EXAMPLE 1:

An electric light and a water pump are located 396 feet from a voltage source of 220 volts. The water pump requires 220 V $\pm$ 5% and 8 amperes of current to operate satisfactorily. The light requires 220 V and 2 amperes for its operation. Number 14 copper wire has been strung between the water pump and voltage source. Determine the voltage that will be impressed across the light and motor. Assume that the currents remain constant. If any problems exist, discuss the possible remedies.

SOLUTION: Follow the step-by-step procedure previously outlined.
1. Given: Source Voltage = 220 V
   Motor Voltage and Currents = 220 V $\pm$ 5%, 8A
   Lamp Voltage and Current = 220 V, 2A
   Wire Size and Material = No. 14, Copper Material
   Distance of Run = 396 feet
2. What voltage is impressed across the motor and lamp? Discuss any problems that might exist and their possible solutions.
3. Draw the schematic. See Figure 6-1.
4. Insert current arrows.
5. Section 1 - Series,   Section 2 - Parallel
6. (a) Lumped resistance of wire can be determined from wire tables or Equation 2-2. From the wire

tables: 396 feet of No. 14 copper wire has a resistance of 1 ohm. Since the total length of the wire is twice the distance of the run, there is a total or "lumped" resistance of 2 ohms for the wire.

(b) Since 5% of 220 V is 11 volts, the minimum motor voltage is 220 − 11 or 209 V.

(c) Conclusion: if the source is 220 V and the minimum motor voltage is 209 V, then no more than 11 volts can be dropped in the wire.

7. Since Section 1 is a series section and all information has now been determined, then:

$$E_{wire} = I_{wire}R_{wire} = 10 \text{ A} \times 2 \, \Omega = 20 \text{ V}$$

CONCLUSION: Number 14 wire has too high a resistance. Number 10 wire would be recommended as a replacement for the No. 14, since it will only drop 7.92 volts.

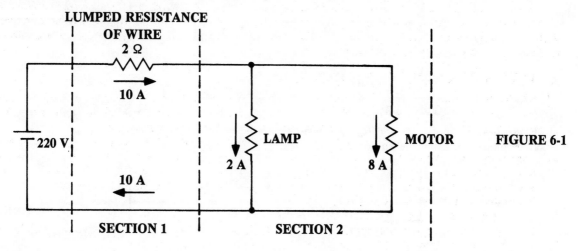

FIGURE 6-1

EXAMPLE 2:

Section the following schematic.

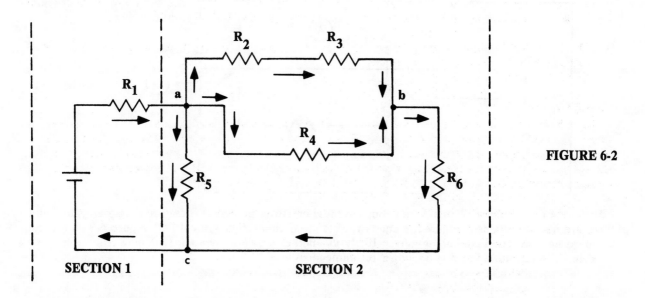

FIGURE 6-2

SOLUTION: Notice that all the circuit current flows through R_1; this is a series section. Section 2 is the major parallel circuit. Resistance R_2 and R_3 are in series and they are both in parallel with R_4. This combination is also in series with R_6. The group consisting of R_2, R_3, R_4, and R_6 is in parallel with R_5.

It should be pointed out that the schematic layout influences the degree of difficulty in analysing the circuit.

CIRCUIT CONNECTION POINTS

Since schematic drawing is an integral part of the technician's duty, several methods of making connections will now be shown. Notice that Node **a** in Figure 6-2, see Figure 6-3, has a circular point to indicate that a connection has been made. Figure 6-4 illustrates another way of making the same connection.

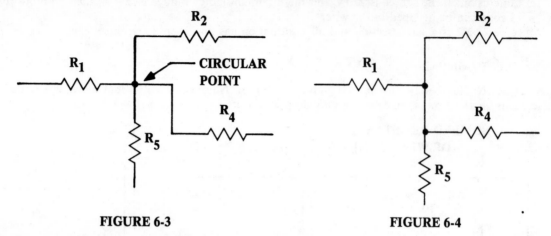

FIGURE 6-3   FIGURE 6-4

In Figure 6-4, the lines of R_1 and R_4 have been separated as shown in Figure 6-5. Figure 6-6 illustrates the same connection using the no-dot method.

FIGURE 6-5   FIGURE 6-6
CONNECTION   CONNECTION

The need for a separation of lines, as shown in Figure 6-6, becomes quite obvious when the schematic of Figure 6-7 is analyzed.

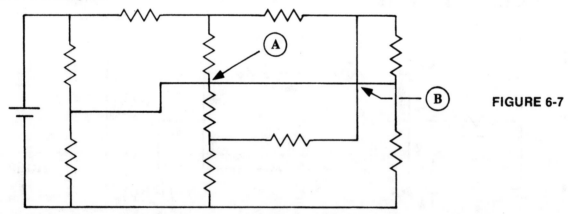

FIGURE 6-7

Is there a connection at Points A and B or are these crossover points? (These are crossover points and there are no connections.) Figure 6-8 shows the standard **connection symbol** and Figure 6-9 shows the standard **no connection symbol**. Figure 6-10 shows another crossover symbol, that may still be encountered in old schematics, but it is no longer recommended.

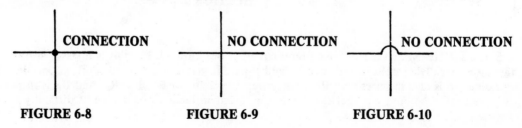

CONNECTION   NO CONNECTION   NO CONNECTION

FIGURE 6-8   FIGURE 6-9   FIGURE 6-10

SUGGESTED OBSERVATIONS

1. Prior to placing a meter into a circuit, estimate the circuit conditions and then set the meter function and range to satisfy these anticipated conditions.
2. Make schematic drawings with a straight edge and a template.
3. Remember that resistance measurements are made with all voltage sources removed.
4. Remove the power from the experiment before making circuit changes.
5. After the variable power supply (VPS) has been connected to the circuit, set it to the desired voltage.
6. Record all measurements and calculations in the data table.

LIST OF MATERIALS

1. VOM
2. VPS
3. Test Circuit: **Consult Your Instructor**
4. Resistors—all at least ½ watt

390 Ω	2.7 kΩ	5.6 kΩ
1 kΩ	3.3 kΩ	6.8 kΩ
1.2 kΩ	3.9 kΩ	10 kΩ
2.2 kΩ	4.7 kΩ	22 kΩ

Alternate Materials:

_____

_____

_____

_____

EXPERIMENTAL PROCEDURE

Emphasis in this experiment will be on troubleshooting, schematic drawing, the sectioning of the drawing for the purpose of circuit analysis, and continued application of current arrows and voltage loops. This experiment also assumes less responsibility for the explicit instructions. That is to say, current ranges, voltage ranges, etc. will not be given. The responsibility of determining these parameters will be yours.

SECTION A: CURRENT FLOW IN A COMPLEX CIRCUIT

1. In the space provided in the data tables, draw the schematic of Figure 6-11.

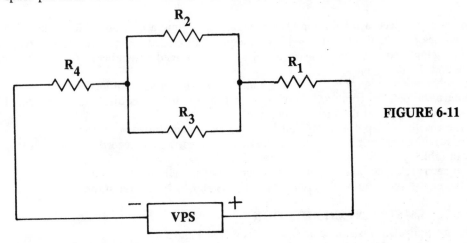

FIGURE 6-11

2. On the schematic, use current arrows to indicate the direction of conventional current flow through each resistance.
3. Determine which components are in series and which components are in parallel.
4. Repeat Steps 1 and 2 for Figure 6-12.

SECTION B: EQUIVALENT CIRCUIT

As stated in an earlier experiment, one circuit is equivalent to another circuit when the circuit parameters, R_T, I_T, E_T are the same. In this part of the experiment, the source voltage will be held constant and the total current will be metered.

1. Construct the circuit of Figure 6-13 and supply it with 30 V from the VPS.
2. Place a milliammeter into the circuit at Point **a**. Be sure the meter is set up properly prior to inserting

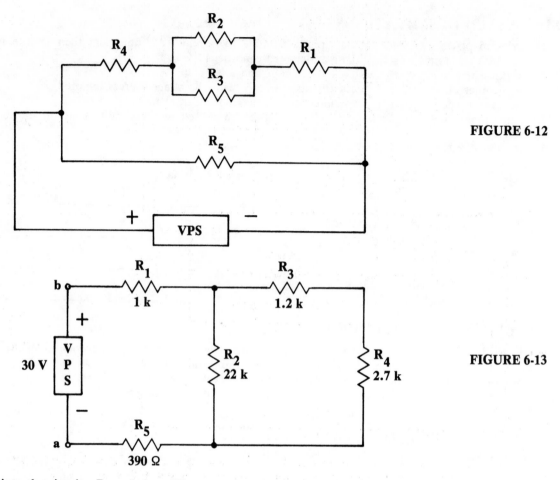

FIGURE 6-12

FIGURE 6-13

it into the circuit. Record the milliammeter reading each time a portion of the circuit is changed.

3. Begin to reduce the circuit of Figure 6-13 to an equivalent circuit by removing series resistor R_3 and R_4 and replacing them with one resistor equal to the combined resistance. Record the value of this resistance and the meter indication.
4. Draw the schematic of Figure 6-13 with R_3 and R_4 now shown as combined resistance, R_{eq-1}.
5. Next, replace the parallel combination of R_2 and R_{eq-1} with one equivalent resistance. Record the value of the resistance and the meter indication.
6. Redraw the circuit of Step 4 with R_2 and R_{eq-1} shown as combined resistance, R_{eq-2}.
7. Finally, reduce the remaining network to a single resistance, R_{eq}. Record the value of this resistance in the data table.
8. Has the meter reading changed during the reduction of the circuit? Explain why.
9. Redraw the circuit of Step 6 with R_1, R_5, and R_{eq-2} shown as one resistance, R_{eq}.

SECTION C: SCHEMATIC INTERPRETATION

1. Calculate the equivalent resistance across Points **a-b** of Figure 6-14.
2. Construct the circuit of Figure 6-14.
3. Measure the total resistance of the network by placing an ohmmeter across Points **a** and **b**. Have your instructor verify this reading.
4. Redraw Figure 6-14 to appear in a more simplified form.

SECTION D: TROUBLESHOOTING

A specially constructed test circuit is used in this part of the experiment. This circuit has been purposely miswired to simulate a malfunction that occurred after nine months of continuous and satisfactory operation. Consider the circuit shown in Figure 6-15 to be the original. Using a VOM, determine which component failed.

1. Compute the voltage E_{ba}, E_{cb}, E_{dc} of Figure 6-15.
2. Supply the test circuit with 24 volts and measure E_{ba}, E_{cb}, and E_{dc} and the current supplied by the voltage source.

50

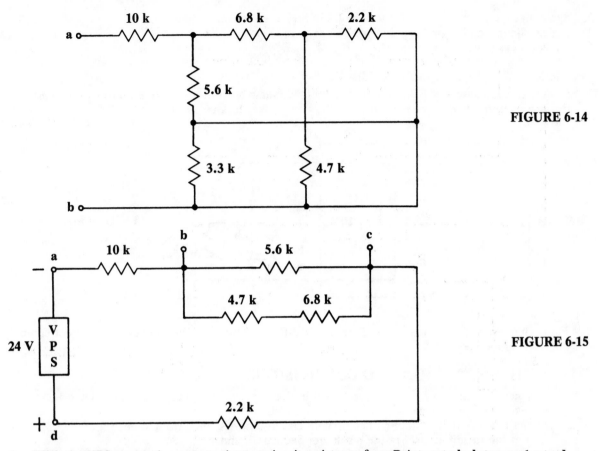

FIGURE 6-14

FIGURE 6-15

3. With the VPS removed, compute the test circuit resistance from Points **a** to **b**, **b** to **c** and **c** to **d**.
4. With the VPS removed, measure the resistance from Points **a** to **b**, **b** to **c**, and **c** to **d**.
5. The data acquired in Steps 1 through 4 provides sufficient information to isolate the defective component. Draw a schematic that indicates which component is defective. Include all working components and their values.

SECTION E: VOLTAGE LOOPS FOR COMPLEX CIRCUITS

Since the algebraic sum of the voltages around any closed loop will equal zero, then the algebraic sum of the voltages around any of the loops in Figure 6-16 will equal zero. Three of several closed loops are shown in Figure 6-16.

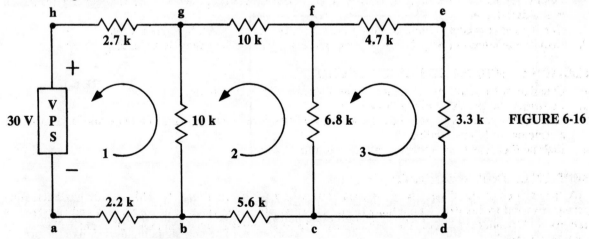

FIGURE 6-16

1. Construct the circuit of Figure 6-16.
2. Measure the voltages around Loop 1 (E_{ba}, E_{gb}, E_{hg}, E_{ah}). Record the answers in the form of an equation equal to zero.

51

3. Show that the algebraic sum of the voltages in Step 2 are equal to zero.
4. Repeat Steps 2 and 3 for Loop 2 (E_{cb}, E_{fc}, E_{gf}, E_{bg}).
5. Repeat Steps 2 and 3 for Loop 3 (E_{dc}, E_{ed}, E_{fe}, E_{cf}).

SECTION F: NODES IN COMPLEX CIRCUITS

1. For the three labeled nodes (**a**, **b**, **c**) of Figure 6-17, show by both calculation and measurement that the current entering each node is equal to the current leaving that node.

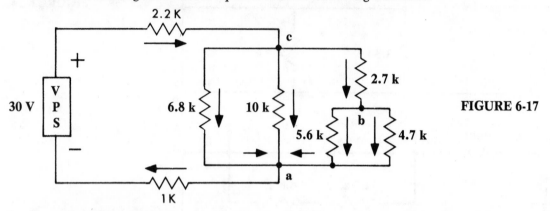

FIGURE 6-17

2. For each of the nodes in Step 1, calculate the % difference between the values of the current entering and leaving the nodes.

DATA INTERPRETATION AND CONCLUSIONS

Write a general summary of the ideas presented in this experiment. This discussion should include:
1. The concept of an equivalent circuit.
2. The need for an understanding of compound circuit analysis when troubleshooting.
3. Account for any differences in the results for Sections E and F.
4. Your own conclusions.

APPLICATIONS

Most electrical and electronic devices utilize the combination of series and parallel circuits.

PROBLEMS

1. Briefly describe what additional information must be known about the 6 volt radio shown in Figure 6-18 to determine the value of the dropping resistor.

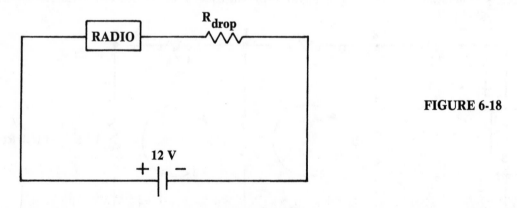

FIGURE 6-18

2. The line going to the signaling device in the Experiment 4 problem has been repaired. It is now desired to place an additional signaling device halfway between the mill and the logging camp. The connecting wire is still No. 14, with 700 ohms of resistance. The new signaling device, like the original one is 800 ohms and requires 10 mA for operation. Determine what modifications must be made to the existing circuit in order for it to work with the additional device. What source voltage is required? Draw a fully labeled schematic of the new signaling circuit.

7 | METER MULTIPLIERS AND SHUNTS VOM LOADING EFFECTS

OBJECTIVES

The insertion of an analog meter, such as the VOM, into a circuit introduces error. The error is caused by current passing through the internal resistance of the basic meter movement and its associated **shunt** and **multiplier** circuits. In this experiment, basic design criteria are given for extending the range of a meter movement through the use of shunts or multipliers. You will learn to determine the internal resistance of a meter movement and to observe the effects of inserting a meter into a circuit. You will also learn to compensate for erroneous readings caused by meter loading.

THE BASIC METER MOVEMENT

The VOM is an extremely versatile test instrument. It is versatile because it can be used to measure a variety of circuit conditions. Also, for each of its functions (voltage, resistance, and current), there are several ranges available. This measuring capability is made possible through internal circuitry that extends the range of the basic meter movement of the VOM. Although this internal circuitry is too complex to be completely discussed at this time, the circuits and design factors used by manufacturers to increase the voltage and current ranges of basic meter movements can be discussed.

Two factors that must be known before any meter movement can have its range increased are:

(1) the manufacturer's rating of the full scale deflection current. A marking of "f.s. = 1mA", or "f.s. = 10 mA", etc. is sometimes printed on the lowest visible portion of the meter scale. If this information is not available from the meter scale or manufacturer specification and data sheets, it can be determined by direct measurement

(2) the internal resistance of the meter. (This information must be determined from manufacturer specifications or measured with special circuitry.) Figure 7-1 shows a basic meter movement symbol modified for use in circuit analysis.

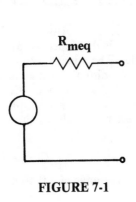

FIGURE 7-1

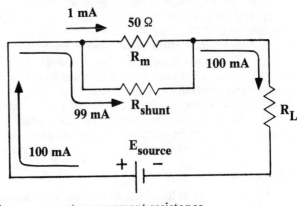

R_m = meter movement resistance
R_{shunt} = meter shunt resistance
R_L = load resistance   **FIGURE 7-2**

CHANGING THE BASIC METER MOVEMENT TO AN AMMETER

Essentially, every meter movement is an ammeter and its range is increased through a process called **shunting**. See Figure 7-2. Observe that the circuit current is 100 mA. Also, note that the 1 mA meter has 99 mA **shunted** around it. Obviously the process of shunting the meter is nothing more than the concepts of the parallel circuit put into practice. Therefore, to effectively design a shunt for a meter movement, the step-by-step procedures set up in Experiment 6 for circuit analysis will be followed.

Example 1 illustrates the technique needed to determine the value of the shunt resistor.

EXAMPLE 1:

Determine the value of the shunt resistance needed to increase the range of a 50 Ω - 1 mA meter to 100 mA. (See Figure 7-2.)

SOLUTION: Since R_m and R_{shunt} are in parallel, the solution is obtained through Ohm's law for parallel circuits.

$$R_m = 50 \; \Omega \qquad\qquad I_m = 1 \text{ mA} \qquad\qquad I_{shunt} = 99 \text{ mA}$$

$$E_{shunt} = E_m = I_m \times R_m = 1 \text{ mA} \times 50 \; \Omega = 50 \text{ mV}$$

$$R_{shunt} = \frac{E_{shunt}}{I_{shunt}} = \frac{50 \text{ mV}}{99 \text{ mA}} = 0.505 \; \Omega$$

where: R_m = meter resistance
R_{shunt} = shunt resistance
E_m = voltage dropped across meter
E_{shunt} = voltage dropped across the shunt
I_m = full scale current of the meter
I_{shunt} = current through shunt

The following Equation, Equation 7-1, has been developed from Example 1. Thus:

$$R_{shunt} = \frac{I_m \times R_m}{I_{shunt}}$$

EQUATION 7-1

A summary of Example 1 clearly indicates that the solution of R_{shunt} requires a knowledge of I_m and R_m. If more than one extension of the basic range is desired, a separate shunt must be designed for each new range.

CHANGING A BASIC METER MOVEMENT TO A VOLTMETER

With the exception of the electrostatic voltmeter, all analog voltmeters are actually current meters that have been graduated for voltage. The conversion of the current meter to a voltmeter requires the relatively simple task of inserting an appropriate resistor.

A re-evaluation of the meter used in Example 1 shows that it has a 50 ohm - 1 mA movement. According to Ohm's law, if 1 mA is flowing through a 50 Ω load, then 50 mV is developed across the resistance. Now suppose a voltage source adjustable from 0 to 50 mV is available. The meter could then be calibrated directly in millivolts and used as a voltmeter for the power supply. Admittedly, it may be difficult to obtain such a power supply and a 50 mV voltmeter would have very limited application. Consequently, it is desirable to extend the range of the meter. Since current is really the limiting factor, design considerations must include some type of a current limiting device. Example 2 shows the steps required to determine the value of the resistor used to limit the meter current.

EXAMPLE 2:

Determine the value of the multiplier resistance needed to increase the range of a 0-50 mV (50 Ω - 1 mA) meter to 0-1 volt.

SOLUTION: The circuit should first be sketched. See Figure 7-3.

Notice that the multiplier resistor, R_{mult} is placed in series with the meter and that it must limit the current to the full scale value by dropping the excess voltage. In this circuit, the resistance of the meter movement will only drop 50 mV; current in excess of 1 mA may damage the meter. The voltage dropped by the multiplier, 950 mV, is the difference between the voltage to be measured, 1 volt (1000 mV) and the 50 mV the meter can safely measure.

Since R_{mult} and R_m are in series, the solution is based on series circuit concepts. Therefore:

$$E_m = I_m \times R_m = 1 \text{ mA} \times 50 \; \Omega = 50 \text{ mV}$$

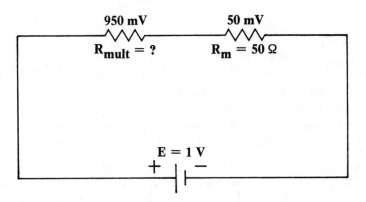

FIGURE 7-3

$E_{mult} = E - E_m = 1000 \text{ mV} - 50 \text{ mV} = 950 \text{ mV}$

$R_{mult} = \dfrac{E_{mult}}{I_{mult}} = \dfrac{950 \text{ mV}}{1 \text{ mA}} = 950 \text{ }\Omega$

where: E = voltage to be measured
E_m = voltage drop across meter
E_{mult} = voltage drop across the multiplier
I_m = full scale current through the meter
R_m = meter resistance
R_{mult} = multiplier resistance

Equation 7-2 can considerably reduce the effort in determining R_{mult}.

$$R_{mult} = \dfrac{E}{I_m} - R_m$$ EQUATION 7-2

As can be seen, shunt calculations require parallel circuit laws for solution, while multiplier calculations require series circuit laws for solution. The solution for the shunt or multiplier can be readily found once the full scale current and internal resistance values of the meter are known.

VOM LOADING EFFECTS

Now that the principles of meter shunts and multipliers are understood, some of the problems encountered when using the VOM must be considered. Whether the VOM is used as a voltmeter or a milliammeter, a certain amount of inaccuracy is introduced by the meter resistance (R_m) and the associated shunt and multiplier resistances. This inaccuracy is called the loading effect. The calculation of the loading effect of a voltmeter is different from that of the ammeter, so each will be considered separately.

AMMETER LOADING

The loading of the circuit, caused by the insertion of a milliammeter, can be calculated when the **current meter sensitivity** is known. Current sensitivity of the meter movement is defined as the amount of current required for full scale deflection. In addition, in order to relate this sensitivity to the loading effect, the voltage necessary for the full scale deflection must be known. This specification may be obtained from the operator's manual or an industrial parts catalog. It can, if necessary, be determined by special measurement. (This measurement technique is developed in Section A of this experiment.)

As an illustration of the meaning of the ammeter sensitivity specification, suppose that a VOM manufacturer advertises his meter to have a full scale sensitivity of 50 μA at 100 mV, 1 mA at 250 mV, and 10 mA at 250 mV. From these specifications, the meter's equivalent resistance (R_{meq} - the meter movement resistance including any shunt) can be calculated. Example 3 uses these specifications to determine the amount of resistance that is introduced in the circuit by the insertion of the meter.

EXAMPLE 3:

What is the equivalent resistance of a milliammeter when the range selector is set of 1 mA and the meter sensitivity is given as 1 mA at 250 mV?

55

SOLUTION:

$$R_{meq} = \frac{E_m}{I_m} = \frac{250 \text{ mV}}{1 \text{ mA}} = 250 \text{ }\Omega$$

where: E_m = voltage developed across the VOM

I_m = current flowing through the VOM

R_{meq} = equivalent resistance that represents the internal resistance of the VOM. This includes the meter, and/or shunt, and series resistance.

Observe that Figure 7-4(a) shows a 1 kΩ load in series with a meter and a 1 volt source. By Ohm's law a current of 1 mA would be expected to flow. However, had the schematic been drawn properly in the first place, see Figure 7-4(b), it would have been obvious that the meter would indicate some value less than the desired 1 mA. After inserting the 250 ohms determined in Example 3, it can readily be seen that the battery is no longer supplying energy to a 1 kΩ resistor, but rather to a 1.25 kΩ resistor. Thus, this reading constitutes error. The amount of error caused by the meter equivalent resistance of 250 ohms is determined in Example 4.

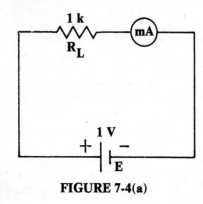

FIGURE 7-4(a)

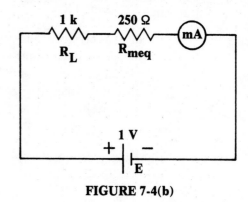

FIGURE 7-4(b)

EXAMPLE 4:

How much error is introduced in a series circuit consisting of a 1 volt source and a 1 kΩ resistor when the current is read with the meter used in Example 3? See Figure 7-4(a) and (b).

SOLUTION:

$$I_{cal} = \frac{E}{R_L} = \frac{1 \text{ V}}{1 \text{ k}\Omega} = 1 \text{ mA}$$

$$I_{meas} = \frac{E}{R_L + R_{meq}} = \frac{1 \text{ V}}{1.25 \text{ k}\Omega} = 0.8 \text{ mA}$$

By Equation 4-7: % error = $\frac{0.8 \text{ mA} - 1.0 \text{ mA}}{1.0 \text{ mA}} \times 100 = -20\%$

As was seen, an R_{meq} of 250 ohms was sufficient to cause a 20% error for this circuit. Generally, if R_{meq} is 1/10th, or less, than the circuit resistance, the error is considered to be negligible.

NOTE: This error is not to be confused with the meter errors discussed in Experiment 1. Also, remember that every meter is merely a resistor and may, therefore, appreciably upset the original circuit when it is inserted to measure the circuit current.

EXAMPLE 5:

What is the microammeter loading effect for a meter sensitivity of 1 μA at 300 mV? See Figure 7-5.

SOLUTION: $I_{cal} = 1 \text{ }\mu A$

$$R_{meq} = 300 \text{ mV}/1 \text{ }\mu A = 300 \text{ k}\Omega$$

56

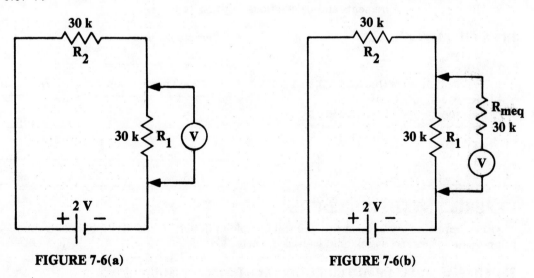

$$I_{meas} = 10\ V/10.3\ M\Omega = 0.97\ \mu A$$

$$\%\ error = \frac{0.97\ \mu A - 1.0\ \mu A}{1.0\ \mu A} \times 100 = -3\ \%$$

FIGURE 7-5

VOLTMETER LOADING

An error is also introduced by the paralleling effect when making a voltage measurement. The magnitude of this error is determined by the voltmeter's sensitivity. The method of expressing this sensitivity is in ohms-per-volt.

$$Ohms\text{-per-volt} = \Omega/V = \frac{1}{I_m} \qquad\qquad \textbf{EQUATION 7-3}$$

where: I_m = full scale current of the basic meter movement

Equation 7-3 states that if a meter requires 50 microamperes to deflect the needle full scale, then, using 1 volt as the **standard** to be measured, the meter must have an equivalent resistance of 20 kΩ if the meter is to be protected. A ratio can be set up between voltage and resistance as shown.

$$\frac{R}{E} = \frac{200\ \Omega}{10\ mV} = \frac{2\ k\Omega}{100\ mV} = \frac{20\ k\Omega}{1\ V} = \frac{1\ M\Omega}{50\ V} = etc. = 20\ k\Omega/V.$$

After carefully examining the ratio concept, it becomes obvious that the ohms-per-volt sensitivity remained constant for all voltage ranges, but the equivalent meter resistance did not. This condition exists because the equivalent meter resistance is determined by multiplying the ohms-per-volt sensitivity (a number) by the range selector voltage setting. For example, the R_{meq} of a 20,000 Ω/V meter set to the 10 V range would result in a R_{meq} of 200 kΩ (20,000 $\Omega \times 10 = 200$ kΩ).

Figure 7-6(a) shows a circuit condition where it is desired to measure the voltage developed across R_1. An evaluation of circuit conditions indicates that a 1 volt drop should appear across each resistor. If the input resistance of the meter, as shown in Figure 7-6(b), is given as 30 kΩ will the meter read 1 volt? No! It will read 0.67 V.

FIGURE 7-6(a)   **FIGURE 7-6(b)**

EXAMPLE 6:

How can the ohms-per-volt rating be used to determine the loading effect of a 20,000 ohms-per-volt

voltmeter set on the 1.5 volt range? See Figure 7-6(a) and (b).
  SOLUTION: Calculation of E_1 has been determined by Equation 4-5. Therefore,

$$E_1 = 1 \text{ V}$$

Figure 7-6(b) has taken on the form of a compound circuit which suggests that R_1 and R_{meq} must be changed to an equivalent value.

$$R_1' = \frac{R_1 \times R_{meq}}{R_1 + R_{meq}}$$

  where: $R_{meq} = 20,000 \text{ }\Omega/\text{V} \times 1.5 \text{ V} = 30 \text{ k}\Omega$

$$R_1' = \frac{30 \text{ k}\Omega \times 30 \text{ k}\Omega}{30 \text{ k}\Omega + 30 \text{ k}\Omega} = \frac{900 \text{ k}\Omega}{60 \text{ k}\Omega} = 15 \text{ k}\Omega$$

and the meter reads: $E = \dfrac{2 \times 15 \text{ k}\Omega}{45 \text{ k}\Omega} = 0.67 \text{ V}$

$$\% \text{ error} = \frac{0.67 \text{ V} - 1.0 \text{ V}}{1.0 \text{ V}} \times 100 = -33 \%$$

This discussion has led to the conclusion that the voltmeter loading effect has to be reduced if the percentage of error is to be reduced. This can be accomplished by using a voltmeter with a higher ohms-per-volt rating. Suppose that the 20,000 ohms-per-volt meter is replaced with a 1,000,000 ohms-per-volt meter in Figure 7-6(b). A quick calculation shows that the percentage of error due to loading has been reduced to -1.47%. The logical conclusion, then, is that if the ohms-per-volt sensitivity is increased, the voltmeter loading effect will be decreased.

SUGGESTED OBSERVATIONS

1. Observe meter polarity when connecting the meter into the circuit.
2. Recall that a voltmeter is an ammeter that has been designed to read voltage.
3. Remember that the VOM, whether used as a voltmeter or an ammeter, will load the circuit and introduce some error.
4. **DO NOT ATTEMPT TO MEASURE THE RESISTANCE OF AN AMMETER WITH AN OHMMETER.**
5. Record all measurements and calculations in the data table.

LIST OF MATERIALS

Alternate Materials

1. VOM
2. VPS
3. 3 potentiometers (100 Ω, 1 kΩ, and 1 MΩ)   * 10 kΩ pot (consult Figure 7-7)
4. Resistors: all at least ½ watt

1 - 100 Ω	1 - 1.5 kΩ
2 - 120 Ω	1 - 3.3 kΩ
1 - 150 Ω	1 - 6.8 kΩ
1 - 220 Ω	1 - 15 kΩ
2 - 1 kΩ	2 - 100 kΩ

EXPERIMENTAL PROCEDURE

After completing this experiment, you will be able to design simple multipliers and shunts and correctly compensate for erroneous readings caused by meter loading.

SECTION A: MILLIAMMETER EQUIVALENT RESISTANCE

1. Set the VOM to measure 1 mA. This is a typical value. However, if the meter being used does not have this range, then use the 100 μA range.

2. Use the following circuit to determine the equivalent resistance of the meter.

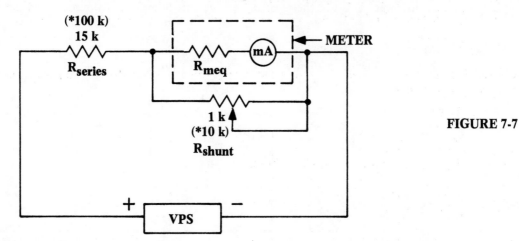

FIGURE 7-7

* These values are used with the 100 μA range setting.

3. With R_{shunt}(1 kΩ potentiometer) out of the circuit, connect the meter in series with a 15 kΩ resistor and the VPS, as shown in Figure 7-7. Slowly increase the supply voltage until the meter reads full scale. Now connect the potentiometer (R_{shunt}) in parallel with the meter. Do not disturb the power supply setting.

4. Adjust the **pot** (potentiometer) until the meter reads half scale. Shut off the supply, remove the pot, and without disturbing the pot setting, measure its resistance. This resistance is equal to the milliammeter equivalent resistance, R_{meq}.

5. Using the equivalent meter resistance, just determined in Step 4 and the full scale current (I_{fs}), write the current specification for this range setting. For example, a meter that has a 10 mA movement and an internal resistance of 5 ohms will have 50 mV developed across it (50 mV = 10 mA × 5 Ω). The current specification for this meter would be written as 10 mA at 50 mV.

6. Using the current sensitivity rating obtained from the operator's manual, or from the instructor, calculate the milliammeter equivalent resistance.

7. Calculate the percent difference between the calculated and measured meter resistance, Steps 6 and 4.

8. Repeat this procedure for the 10 mA meter range. The value of R_{series}, in Figure 7-7, must be reduced to 1.5 kΩ. Similarly, the pot value must be reduced by a factor of 10.

SECTION B: MILLIAMMETER LOADING

1. Using the values given in Figure 7-8, calculate for the **indicated** branch currents leaving nodes **a, b,** and **c.**

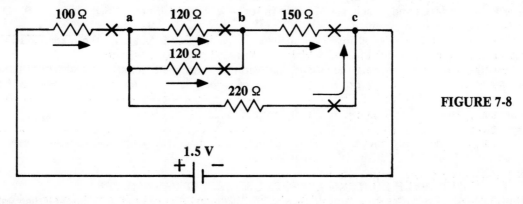

FIGURE 7-8

2. Without taking into account the meter equivalent resistance R_{meq}, measure the current indicated by x's, which enter **a, b,** and **c.**

3. Using the values obtained in Steps 1 and 2, compute the percent difference between computed and measured values.

4. Redraw the circuit and include the equivalent meter resistance introduced at each node. (Three schematics required.)

5. Recalculate the current entering nodes **a**, **b**, and **c** using the new schematics (Step 4).
6. Compute the percent difference between the recalculated current, Step 5, and the measured current, Step 2.

SECTION C: MILLIAMMETER SHUNTS
1. Calculate the current that is flowing in Figure 7-9.

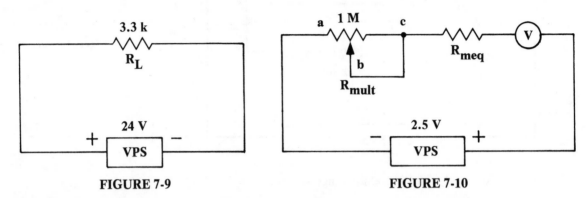

FIGURE 7-9   **FIGURE 7-10**

2. Set the VOM to measure 1 mA. (If this range is not available, then see Section A, Step 1.)
3. Using the equivalent resistance (R_{meq}), determined in Section A - Step 4, and Equation 7-1, calculate and build the shunt needed to expand the meter from 1 mA full scale to 10 mA full scale. See Example 1.

NOTE: $R_{meq} = R_m$. The shunt may have to be built out of several fixed resistors. Have the instructor check your calculations.

4. Connect this shunt across the meter.
5. Construct the circuit of Figure 7-9 and measure the current using the shunted meter of Step 4.
6. Compute the percent difference between the calculated and measured currents (Steps 1 and 5).
7. Increase the value of R_L to 6.8 kΩ. Maintain a source voltage of 24 V. Repeat Steps 1 through 6 for a shunt design of 5 mA full scale.

SECTION D: VOLTMETER EQUIVALENT RESISTANCE
1. Set the VOM to measure 2.5 V DC. (The ranges given are typical. If the meter being used does not have this range, then set it to the nearest range.) Connect the common lead to the negative side of the power supply.
2. With the supply turned completely down, touch the positive test probe to the positive side of the power supply. Slowly increase the supply voltage until the meter reads full scale.
3. Remove the test probe and attach a 1 MΩ pot in series with the voltmeter. See Figure 7-10. Do not disturb the power supply setting.
4. Again touch the probe to the positive side of the supply. Adjust the pot until the meter deflects half scale.
5. Turn off the supply. Disconnect the circuit, and without moving the pot setting, measure the potentiometer resistance between Points **a** and **b**. This resistance is equal to the meter input resistance (R_{mult}).
6. Using the voltmeter ohms-per-volt sensitivity rating, calculate the input resistance (R_{mult}) for the range setting.
7. Compute the percent difference between the calculated and measured equivalent resistance.
8. Repeat Steps 1 through 7 for the 10 V range setting.

SECTION E: VOLTMETER LOADING EFFECTS ON CIRCUIT PARAMETERS
1. Construct Figure 7-11. Set the supply voltage to 2 V. Compute the voltage drop across either resistor.
2. Using the VOM, measure the drop across one of the resistors.
3. Calculate the percent difference between the measured and calculated values.
4. Redraw Figure 7-11 using the voltmeter input resistance (R_{meq}) to modify the voltmeter symbol.
5. Using the schematic of Step 4, calculate the new voltage drop across R_1 with R_{meq} in parallel.
6. Compute the percent difference between the meter reading in Step 2 and the calculated voltage in Step 5.

7. Increase the 2 volt power supply setting to 10 volts and repeat Steps 1 through 6.
8. Increase the 10 volt power supply setting to 40 volts and repeat Steps 1 through 6.

SECTION F: VOLTMETER MULTIPLIER
This section shows how a current meter may be used as a voltmeter.

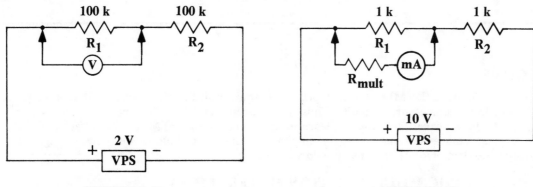

FIGURE 7-11   FIGURE 7-12

1. Construct the circuit in Figure 7-12. Set the source voltage to 10 volts. Calculate the voltage drop across each resistor.
2. Set up the meter to measure 1 mA. Using Equation 7-2, calculate the multiplier needed to convert the 1 mA meter into a 10 volt meter.
3. Using the 1 mA range setting and the multiplier of Step 2, construct a voltmeter with a 10 volt range.
4. Using the meter constructed in Step 3, measure the voltage drop across R_1 and the voltage drop across R_2.
5. Calculate the percent difference between the measured and computed value.
6. Calculate the ohms-per-volt sensitivity of the constructed meter.
7. Repeat Steps 1 through 5 for a source voltage of 28 volts (Step 1) and a multiplier (Step 2) for a 15 volt range.
8. Repeat Steps 1 through 5 for a source voltage of 40 volts and a multiplier for a 25 volt range.

DATA INTERPRETATION AND CONCLUSIONS
Write a general summary of the ideas presented in this experiment. This discussion should include:
1. The possibility of introducing error into circuit measurements when a VOM is used.
2. The meaning of ammeter and voltmeter sensitivity specifications.
3. The need for shunts and multipliers.
4. Your own conclusions.

APPLICATIONS
Among the various devices utilizing shunts and multipliers are: (1) tachometers, (2) fuel gauges, (3) ammeters, (4) power supplies, and (5) relays.

PROBLEMS
1. A circuit consists of 2 resistors (4.7 kΩ and 2.2 kΩ) connected in series with a 10 volt source. Using a 1000 ohms-per-volt meter, measure the voltage across the 4.7 kΩ resistor with the range selector set to 10 volts. What will the meter reading be? Draw a schematic of the circuit using the meter equivalent resistance.
2. Why does the voltmeter loading effect decrease when the range setting is increased?
3. Using an industrial parts catalog, select three different VOM voltmeter sensitivities and list the manufacturer's name and meter model number. Compare the ohms-per-volt rating to the cost of the VOM.

INTRODUCTION TO ELECTRONIC MULTIFUNCTION METERS

OBJECTIVES

The Digital Voltmeter (DVM) and other types of electronic multifunction meters (MFM) take practically no current from the measured circuit. Thus the electronic multifunction meter effectively minimizes loading effect. In this experiment you will learn to use the electronic MFM to take voltage and resistance measurements. Voltage notation is again emphasized and a technique for determining the input resistance of the electronic MFM is also provided.

THE ELECTRONIC MULTIFUNCTION METER (MFM)

As demonstrated in the last experiment, loading effects of the VOM must be reduced if precision measurements are to be obtained from electronic circuits. In order to reduce loading effects, the input resistance of the meter must be increased. Initially, the increase in meter input resistance was accomplished by using the vacuum tube in the development of the Analog Vacuum Tube Voltmeter (VTVM). In recent years solid state technology has replaced the vacuum tube in both the digital and the analog multifunction meter. In fact the digital voltmeter has replaced the analog meter in many measurement applications.

Because both analog and digital solid-state multifunction meters are in use, the term **electronic multifunction meter** (MFM) will be used to identify both types of meters.

SUGGESTED OBSERVATIONS

1. When testing a circuit, handle only one probe at a time. This minimizes the danger of shock and fatal injury.
2. When measuring voltage, it is standard practice to remove the power before connecting the test leads.
3. If the magnitude of the voltage under test is unknown, start on the highest range and work down.
4. Prior to making a resistance measurement, all power to the circuit under test **MUST be removed**.
5. Next to the instructor, the operator's manual is the best source of information about a MFM.
6. Record all measurements and calculations in the data tables.

LIST OF MATERIALS

Alternate Materials

1. Electronic MFM
2. VOM
3. Operator's manual
4. VPS
5. Resistors—all at least ½ watt

1 kΩ	4.7 kΩ	10 kΩ
3.3 kΩ	6.8 kΩ	10 MΩ

_____

_____

_____

_____

EXPERIMENTAL PROCEDURE

The emphasis in this experiment will be on the set-up and operation of the electronic MFM.

SECTION A: RESISTANCE MEASUREMENT

1. Insert the test leads in their proper jacks. (Consult the operator's manual or check with the instructor.) Set the function switch to ohms. Connect the ohms probe to the common test lead and check to see that the meter zeros. (This procedure of connecting the leads is generally referred to as "shorting the leads".)
2. Using six color-coded resistors, read the color code and measure the resistance with the electronic MFM. Calculate the percent difference using Equation 2-1. Enter these answers in the data table. If difficulty is encountered in using the ohmmeter, consult the operator's manual.

SECTION B: DC VOLTAGE MEASUREMENT

1. Set the function switch to +DC. Plug the test leads in the proper jacks. (If in doubt, see the operator's manual.) Initially, set the range switch to measure 10 volts. Refer to Figure 8-1 and measure the following voltage drops:

$$E_{de} \quad\quad E_{ba} \quad\quad E_{ab}$$
$$E_{cd} \quad\quad E_{cb} \quad\quad E_{ed}$$
$$E_{bd} \quad\quad E_{db} \quad\quad E_{bc}$$
$$E_{dc}$$

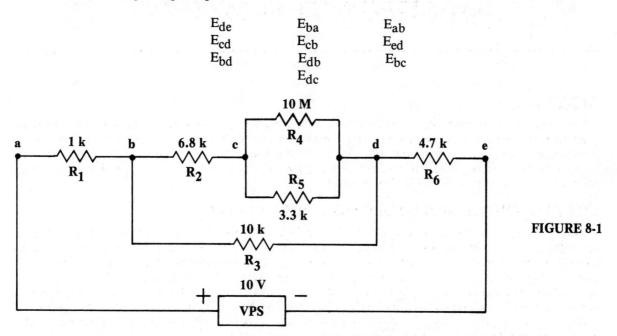

FIGURE 8-1

2. Record the voltage drops in the data table using both magnitude and polarity. If a downward deflection of the analog meter occurs, then the function switch should be changed from +DC to −DC.

SECTION C: DETERMINING CURRENT VALUES WITH THE MFM BY CALCULATION

1. To determine the current entering or leaving node **d** in Figure 8-1, first measure the value of R_6. Using the value of E_{de} (from the data table) and the resistance just measured, compute the value of I flowing through $R_6(I_d = E_{de}/R_6)$.
2. Repeat Step 1 for nodes **c** and **b**. Select the appropriate resistors.
3. Remove the voltage source and measure the total resistance across terminals **a** and **e**. Using the voltage of the source and the resistance just measured, compute I_T.
4. Using Equation 4-7, calculate the percent difference between the current computed in Step 3 (consider this to be a measured value) and the current computed in Step 1 (consider this to be the computed value).
5. Draw a schematic of the equivalent circuit of Figure 8-1. This equivalent circuit must include a voltage source and one resistor. Clearly label all component values.

SECTION D: OHMMETER POTENTIALS

This section is intended to show that the test leads of both the VOM and the analog electronic MFM have a potential across them when their function switches are in the ohms positions.

1. Set up the VOM to measure resistance, using the R × 1 range, and the electronic MFM to measure approximately 5 volts.
2. Connect the common leads of the instruments together, and then touch the remaining probes together. Record both the amplitude and the polarity of the VOM's voltage. Is the black lead positive or negative?

NOTE: Set the MFM to measure 30 V.

3. With the meter set to 30 V, repeat Steps 1 and 2 for all the VOM ohmmeter ranges.

CAUTION: Since it has just been demonstrated that a potential does exist across the test leads of an instrument when it is set up for measuring resistance, it is very important that no attempt is made to use an ohmmeter to measure the resistance of meter movements.

SECTION E: DETERMINING ANALOG MFM EQUIVALENT RESISTANCE

In Experiment 7, the equivalent resistance of the voltmeter was empirically determined. In this section, the electronic MFM input resistance (equivalent resistance) will be determined using Equation 8-1 and Figure 8-2.

$$R_{meq} = \frac{E_2 R_s}{E_1 - E_2}$$

EQUATION 8-1

where: R_{meq} = equivalent internal resistance of the VOM or electronic MFM

R_s = external resistance inserted in series with the electronic MFM test probe

E_1 = meter reading (preferably full scale deflection) without R_s

E_2 = meter reading after the insertion of R_s

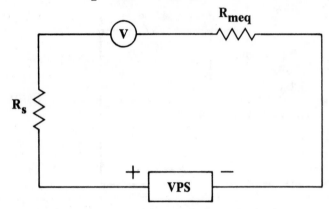

FIGURE 8-2

1. Set the electronic MFM to measure 10 volts DC. Attach the common test lead to the common side of the VPS (minus terminal). With the supply turned completely down, touch the positive DC meter probe to the positive side of the supply. Now, slowly increase the supply voltage until the meter pointer is deflected full scale. Read the meter and record this voltage as E_1.
2. Without moving the power supply setting, remove the positive meter probe and insert R_s in series with meter—as in Figure 8-2. The value of R_s is selected at random. In this case a 10 M resistor was selected. Reconnect the circuit and read the meter. Record the reading as E_2.
3. Using Equation 8-1 and the previously determined values of E_1, E_2, and R_s, calculate for the unknown meter internal resistance (R_{meq}).

NOTE: Because the DVM may have a very high internal resistance, the technique just described may not work. If difficulty is experienced, consult your instructor.

DATA INTERPRETATION AND CONCLUSIONS

Write a general summary of the ideas presented in this experiment. This discussion should include:
1. Safety precautions when using the electronic MFM with live circuits.
2. The number of ohm and voltage ranges of the electronic MFM compared to the VOM.
3. The voltmeter input resistance of the electronic MFM versus that of the VOM.
4. Your own conclusions.

APPLICATIONS

The electronic MFM can be used to make voltage and resistance measurements of almost any electronic devices such as: (1) radios, (2) televisions, (3) HiFi systems, (4) CB equipment, (5) ham radio equipment, (6) intercoms, and (7) public address systems. Because of the high input resistance, the electronic MFM will give more accurate readings.

PROBLEMS

1. Calculate the percent difference between the published input resistance of the electronic analog MFM used in this experiment and that value obtained in Section E.
2. From a parts catalog determine the cost of the electronic MFM used in this experiment. (If this particular electronic MFM cannot be located, determine the cost of a similar one.)

9 || VOLTAGE DIVIDERS

OBJECTIVES

Voltage dividing networks are used in most electrical and electronic equipment. This experiment stresses voltage divider design.

VOLTAGE DIVIDERS

Voltage dividers take on a variety of different forms. These include: (1) several resistors connected in series, (2) single or multi-tapped power resistors, (3) potentiometers, (4) rheostats, or (5) any combination thereof. The final configuration that any divider assumes is limited only by the imagination of the designer. Figure 9-1 shows the acceptable symbols used to designate: (a) the potentiometer, (b) a tapped potentiometer, (c) an adjustable resistor or rheostat, and (d) the fixed-tap resistor.

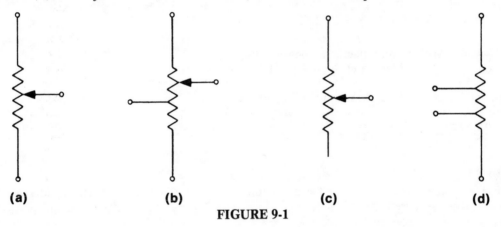

| (a) | (b) | (c) | (d) |

FIGURE 9-1

Potentiometers, variable resistors, and rheostats are available in many sizes and shapes—each designed to do a specific job. There are numerous factors that describe the characteristics of these components. Some of the factors to be discussed in this experiment are: (1) the linear resistive element, (2) the non-linear resistive element, and (3) the power ratings.

LINEAR AND NON-LINEAR RESISTIVE ELEMENTS

A linear resistive element is one that produces equal resistance changes for equal physical changes. For instance, a 2700 ohm potentiometer, or rheostat, having a maximum rotational change of 270 degrees, would have a 10 ohm change for every degree of rotation. For the resistor with a sliding tap, the physical change could be based on the actual length of the resistor and the distance the tap is from one end. As an illustration, assume a resistor has a physical length of 3 inches and a resistance of 300 ohms. If the resistance is linear, then, for every inch of resistor there would be 100 ohms of resistance.

Since non-linearity is built·into the resistive elements in varying degrees, the non-linear element cannot be described with the ease and accuracy of the linear element. When working with non-linear resistive elements, you must consult the manufacturers' specification sheets which contain information on the **taper** (non-linearity) of the element.

POWER RATING

Power ratings are specified for each electronic component that is capable of dissipating power. In Experiment 2, it was pointed out that the physical size of the resistor indicates to some degree the amount of power that can be dissipated. This concept will now be covered in detail.

Figure 9-2 illustrates how a power gradient is established. Assume that the following characteristics describe the resistor. Length: 10 inches, Diameter: 3/4 inch, Resistance: 10 ohms, and Power rating: 10 watts. The power rating of 10 watts means that this is the maximum amount of power that can be safely dissipated over the entire length of 10 inches.

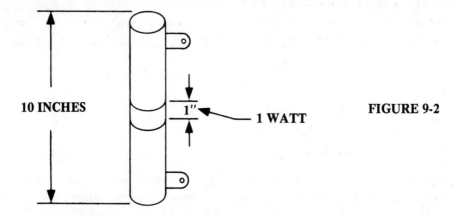

10 INCHES   1" ← **1 WATT**   **FIGURE 9-2**

To establish a power gradient for the resistor, simply divide the power rating by the length.

$$\text{Power Gradient} = \frac{\text{MPR}}{\text{L}}$$ **EQUATION 9-1**

where: MPR = manufacturers' power rating
          L = length of the resistive element in degrees, inches, etc.

Thus, the power gradient for the resistor in Figure 9-2 is:

$$\frac{10 \text{ watts}}{10 \text{ inches}} = 1 \text{ watt/inch}$$

EXAMPLE 1:

A linear 2700 ohm, 50 watt rheostat has a maximum rotation of 270°. How many watts can be safely dissipated if the resistance is known to be approximately 500 ohms after final adjustment?

SOLUTION: First, draw a schematic to get a clear picture of what is taking place. See Figure 9-3 where:

$$\frac{500 \text{ }\Omega}{2700 \text{ }\Omega} = \frac{\text{X}°}{270°} \quad \text{and} \quad \text{X}° = \frac{270 \text{ }\Omega \times 500 \text{ }\Omega}{2700 \text{ }\Omega} = 50°$$

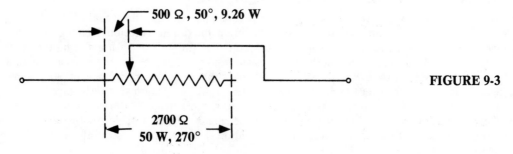

500 Ω , 50°, 9.26 W

2700 Ω
50 W, 270°

FIGURE 9-3

By Equation 9-1, the power gradient is: $\dfrac{50 \text{ watts}}{270 \text{ degrees}} = 0.185 \text{ watts/degree}$

Since this is a linear potentiometer, it becomes evident that: $\dfrac{500 \text{ }\Omega}{2700 \text{ }\Omega} = \dfrac{\text{X}}{50 \text{ W}}$ Thus, X = 9.26 watts.

An alternate solution is obtained through the ohms-per-degrees concept which would be 10 Ω/degree; therefore, the shaft is rotated 50 degrees to obtain 500 ohms. Multiplying 50 degrees by 0.185 watts per degree results in 9.26 watts.

ELECTRIC POWER EQUATIONS

To further explore the power concept, assistance will be required from the **power equations**. These equations express power in terms of voltage, current, and resistance.

(a) $P = IE$ (b) $P = \dfrac{E^2}{R}$ (c) $P = I^2 R$ **EQUATION 9-2**

where: P = power in watts
I = current in amperes
E = voltage in volts
R = resistance in ohms

EXAMPLE 2:'
A linear adjustable 25-ohm, 50-watt resistor is 4 inches long. See Figure 9-4.

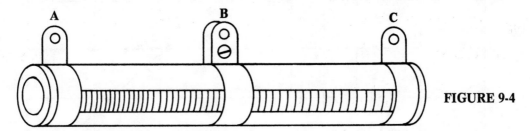

FIGURE 9-4

From this information, determine the following:
(a) The resistance between terminal lug B and lug C when the tap is 1⅓ inches away from lug C.
(b) The maximum power that can be dissipated in this section.
(c) The maximum current that can be safely passed in this section.
(d) The maximum current that can be safely passed through the full 25 ohms.

SOLUTION: Draw a schematic, see Figure 9-5.

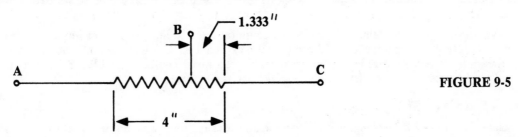

FIGURE 9-5

(a) By ratio: $\dfrac{1.333 \text{ in.}}{4 \text{ inches}} = \dfrac{R_X}{25 \text{ Ω}}$ Thus: $R_X = 8.33 \text{ Ω}$

(b) By Equation 9-1: $\dfrac{50 \text{ W}}{4 \text{ in.}} = \dfrac{12.5 \text{ W}}{1 \text{ in.}} = \dfrac{P_X}{1.333 \text{ in.}}$ Thus: $P_X = 16.66 \text{ watts}$

(c) By rearranging Equation 9-2(c): $I = \sqrt{\dfrac{P}{R}} = \sqrt{\dfrac{16.66 \text{ W}}{8.33 \text{ Ω}}} = \sqrt{2}$ Thus: $I = 1.414 \text{ A}$

(d) By rearranging Equation 9-2(c): $I = \sqrt{\dfrac{P}{R}} = \sqrt{\dfrac{50 \text{W}}{25 \text{ Ω}}} = \sqrt{2}$ Thus: $I = 1.414 \text{ A}$

POWER DERATING

The power rating specified by the manufacturer was obtained after many complex tests were performed. One test determines the maximum power needed to raise the resistor to a temperature that is stipulated in an industrial standard. As an illustration, examination of the curve in Figure 9-6 shows that if the surrounding temperature (ambient temperature) is 150°C, then **absolutely** no power can be dissipated in the resistor. On the other hand, if the ambient temperature is between zero and 70°C, then 100% of the power rating may be used and no derating is required. At 70°C ambient and 100% rating power, the temperature at the resistor **hot spot** (hottest portion of resistor) will be 150°C. If the ambient temperature is 100°C, then 60% of the rated power will cause the resistor temperature to increase 50°C to 150°C. Thus, a total temperature of 150°C would be measured on the surface of the resistor (100°C ambient + 50°C power dissipation = 150°C).

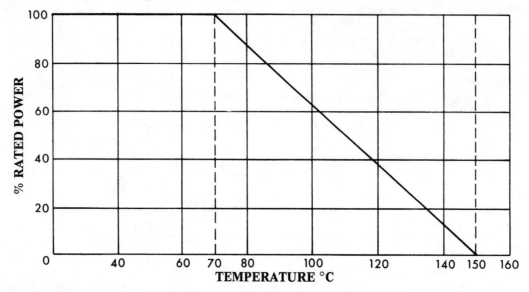

FIGURE 9-6: POWER DERATING CURVE

Since heat is always detrimental to electronic and electrical components, it is unwise to operate resistors at their 100% ratings. Good design requires the use of a power rating from 2 to 4 times greater than the computed wattage. Therefore, if the computed power indicates that a resistor would dissipate 1 watt, good design practice would dictate that a 2 to 4 watt resistor be used. Even a 5 or 10 watt resistor may be selected.

NOTE: The use of the word "ambient" has many connotations. In industry, for example, a motor manufacturer may state that the motor temperature may not exceed a 65°C rise above ambient. In the case of a motor, the ambient temperature is referenced at 40°C. To avoid confusion, **ALWAYS** check the manufacturers' specification sheets to determine the reference temperature.

CIRCUIT ANALYSIS

Associated closely with the engineering technician's design ability is an inherent responsibility for selecting and specifying components. The following example will serve to illustrate this point.

EXAMPLE 3:

The project engineer needs to vary the voltage from a fixed 200 volt source. The circuit shown by the sketch of Figure 9-7 is to be constructed by the technician. Should the circuit be constructed as presented?

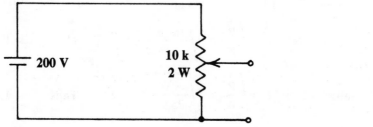

FIGURE 9-7

68

SOLUTION: No! The call-out of a 2-watt potentiometer is an error. By Equation 9-2(b), it can be shown that the potentiometer will be dissipating 4 watts of power. Thus, as part of a team, it is the technician's responsibility to make certain that parts call-outs are correct. Building this circuit as it was originally specified would certainly have resulted in a failure—perhaps during an expensive experiment.

EXAMPLE 4:

What is the minimum value of resistance that the load (R_L) can have without exceeding the current limitation of the 10 kΩ, 2 W pot? See Figure 9-8.

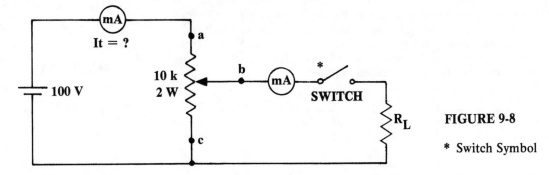

FIGURE 9-8

* Switch Symbol

SOLUTION: By Equation 9-2(c), the maximum current the pot can pass is 14.14 mA.

$$I = \sqrt{\frac{2\,W}{10\,k\Omega}} = 14.14\,mA$$

With the switch in the open position as shown, there is 10 mA of current flowing through the potentiometer. Since the maximum current is 14.14 mA and the current without a load is 10 mA, the maximum current that the load can demand is 4.14 mA. The minimum resistance is calculated on the basis that the potentiometer is set at the maximum voltage (100V). By Ohm's law, R_L = 24,150 ohms.

$$R_L = \frac{100\,V}{4.14\,mA} = 24{,}150\ ohms$$

EXAMPLE 5:

Design a voltage divider that would operate from a 150 volt power source and will provide the following: 150 V at 50 mA and 100 V at 10 mA. A stabilizing or bleeder current of 10 to 20% of the total load current must also flow through the bleeder resistor (R_B).

SOLUTION: Sketch the basic circuit. See Figure 9-9.

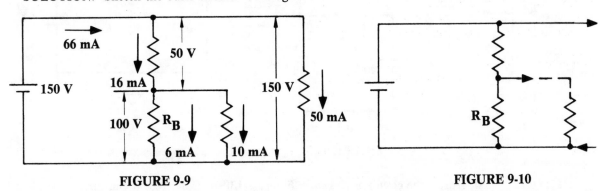

FIGURE 9-9   **FIGURE 9-10**

The 150 V, 50 mA load is not an integral part of the divider and it has been removed as noted in Figure 9-10. The remaining configuration of Figure 9-10 suggests that an adjustable power resistor might be satisfactory. By Ohm's law the resistance required to drop 50 V at 16 mA is 50V/16 mA = 3,125 Ω. The remaining portion of the divider will have only 6 mA, 10% of 60 mA, flowing through it. Thus, its resistance is 100 V/6 mA = 16.67 kΩ. If a single resistor is to be used, the total resistance will be 3.125 kΩ plus 16.67 kΩ or 20.895 kΩ   A 20 kΩ resistor would be satisfactory. The power rating required for the resistor can be evaluated by first determining the maximum amount of current that flows through the resistor—squaring it—and multiplying it by the total resistance. The product is the minimum power rating of the resistor, e.g., 16 mA is the maximum current through the resistor, squared it becomes 256×10^{-6}, multiplied by 20×10^3 results in a minimum power rating of 5.12 watts. However, good design practice would require that the minimum power rating be from 10 to 20 watts.

THE REFERENCE POINT

Beginning with this experiment, the actual symbols of the power supply, or supplies, will no longer be shown. Instead the voltage requirements will be noted as shown in Figure 9-12 and 9-13. Observe that a new schematic symbol has been added, Figure 9-11, which indicates a **common connection** point. Unless otherwise noted, the common connection point is to be considered the reference point. This symbol is sometimes incorrectly called **ground**.

FIGURE 9-11

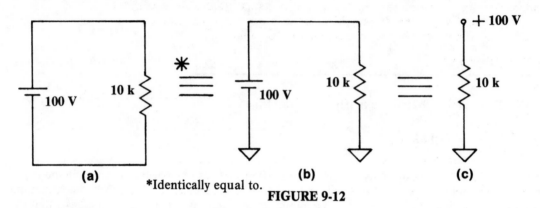

*Identically equal to.
FIGURE 9-12

EXAMPLE 6:

Determine E_{ac}. See Figure 9-13(a).

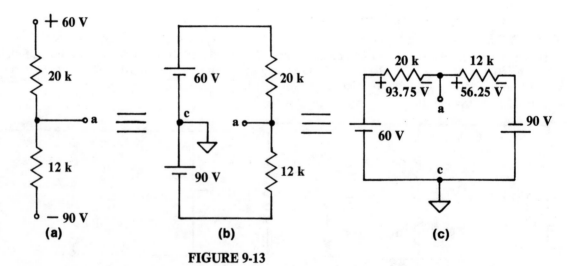

FIGURE 9-13

SOLUTION: First, redraw the circuit as shown in Figure 9-13(b). If desired, it can further be placed in the form of Figure 9-13(c). Using the closed loop concept, it can be seen that the two voltage sources are series aiding, 90 V + 60 V = 150 V. Furthermore, the resistors are also in series, 12 kΩ + 20 kΩ = 32 kΩ. The voltage across each of the resistors can be determined by the voltage divider equation, Equation 4-5.

$$E_{12\,k\Omega} = \frac{150\text{ V} \times 12\text{ k}\Omega}{32\text{ k}\Omega} = 56.25\text{ V} \qquad E_{20\,k\Omega} = \frac{150\text{ V} \times 20\text{ k}\Omega}{32\text{ k}\Omega} = 93.75\text{ V}$$

70

Label all polarities and insert the appropriate voltage drops as shown in Figure 9-13(c). Starting at Point **c** and moving in the CW (clockwise) direction stopping at Point **a** results in a net voltage of:

$$+60 \text{ V} - 93.75 \text{ V} = -33.75 \text{ volts}$$

Moving in the CCW (counter clockwise) direction shows:

$$-90 \text{ V} + 56.75 \text{ V} = -33.75 \text{ volts}$$

Thus, $E_{ac} = -33.75$ volts.

SUGGESTED OBSERVATIONS
1. Continued to draw schematics as neat as possible.
2. Review the methods of noting circuit connection and non-connection presented in Experiment 6.
3. Before attaching the VPS to the circuit, be sure the supply voltage is turned down.
4. Read through each experimental section before doing the experiment.
5. Record all measurements and calculations in the data table.

LIST OF MATERIALS
Alternate Materials

1. MFM
2. VPS
3. Potentiometers, 1 kΩ and 1 MΩ
4. Resistors—all at least ½ watt

1 kΩ	10 kΩ	100 kΩ (two)
3.3 kΩ	15 kΩ (two)	470 kΩ
3.9 kΩ	33 kΩ	1 MΩ
4.7 kΩ	47 kΩ	

EXPERIMENTAL PROCEDURE

In this experiment the emphasis is on circuit analysis, design, and troubleshooting.

SECTION A: POTENTIOMETER RESISTANCE CHARACTERISTICS
1. Using an ohmmeter determine the resistance from **a** to **c** in Figure 9-14. Does rotating the shaft have any effect on the reading?

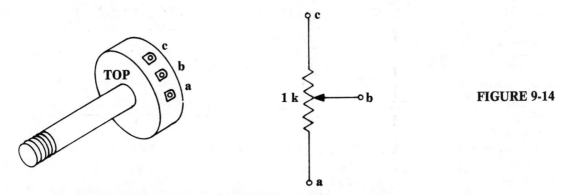

FIGURE 9-14

2. With the ohmmeter across terminals **a** and **b**, complete the following:
    (a) Ohmmeter reading with shaft rotated full CCW (counter-clockwise).
    (b) Ohmmeter reading with shaft rotated full CW (clockwise).
    (c) Does shaft rotation have any effect on the ohmmeter reading?
3. Repeat Step 2 with the ohmmeter across terminal **b** and **c**.

SECTION B: POTENTIOMETER VOLTAGE DIVIDER
1. The source voltage (E_{ac}) in Figure 9-15 must be maintained at 10 volts. E_{bc} is to be adjusted in 1 volt increments starting at 0 volts and increasing to 10 volts. After each adjustment of E_{bc}, E_{ac} is to be

measured. After measuring E_{ac}, remove the source voltage and measure R_{bc} and R_{ab}. (**Make certain that the function switch is changed from ohms to volt PRIOR to the next measurement!**)

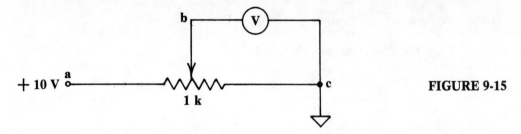

FIGURE 9-15

2. Since the amount of voltage dropped across a resistor varies directly with the amount of resistance, then, the voltage ratio, E_{bc} to E_{ac}, is the same as the resistance ratio, R_{bc} to R_{ac}. Show that

$$\frac{E_{bc}}{E_{ac}} = \frac{R_{bc}}{R_{ac}}$$

is true for values of E_{bc} equal to 2, 5, and 8 volts. Use the data obtained in Step 1 and the measured value of R_{ac} to complete the ratio.

3. As in Step 2, show that the ratio is true for values of E_{ab} equal to 3, 6, and 9 volts.

$$\frac{E_{ab}}{E_{ac}} = \frac{R_{ab}}{R_{ac}}$$

SECTION C: VOLTAGE DIVIDER APPLICATION

Assume a circuit, similar to the television's horizontal hold circuit of Figure 9-16(a), is needed for a particular application. It is desired to limit the maximum as well as the minimum output voltage of the divider. How can this be achieved?

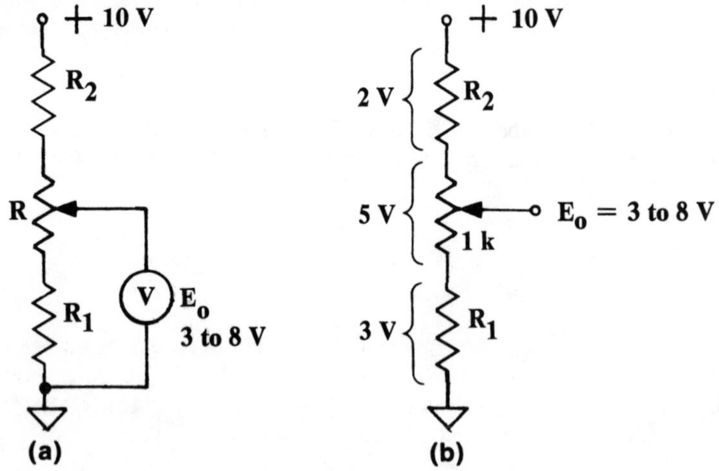

FIGURE 9-16

(a)   **(b)**

SOLUTION: Pick a potentiometer value of 1 kΩ as shown in Figure 9-16(b) and solve for R_T, and R_2.

$$R_T = \frac{E_T}{I_T} = \frac{10}{\dfrac{8-3}{1 \text{ k}\Omega}} = \frac{10}{5 \times 10^{-3}} = 2 \text{ k}\Omega$$

Transposing Equation 4-5 and solving for R_x:

$$R_x = \frac{E_x R}{E_A}$$

Replace the notation of Equation 4-5 to fit that of the voltage divider and solve.

72

For R_1 where $E_1 = 3$ V:
$$R_1 = \frac{E_1 R_T}{E_T} = \frac{3 \times 2 \text{ k}\Omega}{10} = 600 \text{ } \Omega$$

For R_2, where $E_2 = 2$ V:
$$R_2 = \frac{E_2 R_T}{E_T} = \frac{2 \times 2 \text{ k}\Omega}{10} = 400 \text{ } \Omega$$

Since 400 and 600 ohms are not standard carbon resistor values, the needed values could be built up using several resistors or a different type of resistor, such as a wire-wound resistor. Another possibility would be to change the range of E_0 so that lower cost carbon resistors could be used.
1. Using Figure 9-16(a) and a 1 kΩ pot, determine by calculation the value of R_1 and R_2 for an E_0 of 5.5 to 14.5 volts when the supply voltage is 18 V.
2. Using the values determined in Step 1, construct the circuit and check the calculations by measuring the E_0 range.

SECTION D: DESIGN PROBLEM
Prior to shipment, it is desired to replace the 1 kΩ "Factory Adjust" potentiometer of Figure 9-17 with fixed resistors.

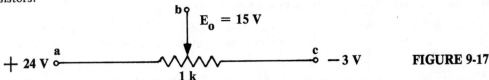

FIGURE 9-17

1. Draw a schematic of Figure 9-17, including the source symbols so the circuit may be easily analyzed.
2. Calculate for the values of two fixed resistors (R_{ab} and R_{cb}) needed to replace the 1 kΩ pot. The output voltage must remain at 15 volts in relation to the common point.
3. Using a derating factor of 3, determine the wattage rating for each resistance.
4. Using a derating factor of 2, what is the wattage rating of the 1 kΩ pot?
5. Construct the circuit of Figure 9-17 and experimentally determine the resistance division of the pot by setting E_0 to 15 V and then, with the voltages removed, measure the resistance R_{ab} and R_{cb}.

DATA INTERPRETATION AND CONCLUSIONS
Write a general summary of the ideas presented in this experiment. This discussion should include:
1. The need for wattage calculation prior to the insertion of the resistance in a circuit.
2. The reason for derating the wattage specifications of resistive components.
3. The need for neat, concise, and neatly labeled schematics.
4. Your own conclusions.

APPLICATIONS
Because of its extreme versatility, the voltage divider is found in most electronic equipment. Although the names may not be familiar, some of the uses for voltage dividers are: (1) volume controls, (2) tone controls, (3) contrast (TV), (4) brightness (TV), (5) color adjust (TV), (6) T-pads, (7) balance controls (stereo), (8) distance control (headlight dimmer), and (9) fixed attenuators.

PROBLEMS
1. Using a parts catalog, select three potentiometers with different power ratings. Determine the cost, physical size, and other significant factors.
2. Using a parts catalog, determine the cost of a "locking" potentiometer.
3. What is the cost of a 10 turn potentiometer?
4. How many types of indicating dials are shown for the 10 turn potentiometer?
5. Describe a trimming potentiometer.
6. Using the information obtained in Steps 2 and 3, Section D, determine the cost of modifying 100 such circuits. In a short paragraph, describe what type of resistors were chosen and why?

10 | THE SUPERPOSITION THEOREM

OBJECTIVES

The superposition theorem provides a method for using Ohm's law in the solution of a multi-source circuit. This experiment demonstrates the superposition theorem.

THE SUPERPOSITION THEOREM

THE SUPERPOSITION THEOREM states: **Any network containing only linear bilateral elements can be reduced to as many individual circuits as there are power sources.**

Each circuit can then be solved by Ohm's law, using one source at a time, with all other sources replaced by their internal resistances. The actual circuit current is then determined by obtaining the algebraic sum of the individual currents.

NOTE: A linear element is defined as one in which equal changes in voltage cause equal changes in current. To be classed as a bilateral element, the same magnitude of current must flow if the polarity is reversed.

Thus, the superposition theorem uses the concept that the original circuit, containing only linear bilateral elements, can be modified for the purpose of analysis. Such modifications take the form of several individual circuits, each containing only one energy source. The removed sources are replaced with their internal resistance of zero ohms (short circuit) if they are voltage sources.

ANALYSIS OF A CIRCUIT CONTAINING TWO SOURCES

Consider the circuit of Figure 10-1. This circuit cannot be solved by Ohm's Law. However, a quick examination of the circuit shows that there are two voltage sources, 14 V and 28 V. According to the superposition theorem, two additional circuits must be drawn, each containing one voltage source. Visualize each circuit as being drawn on a clear plastic sheet. The following steps lead to a successful solution.

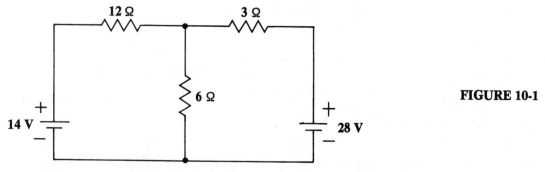

FIGURE 10-1

1. Initially, select one source to work with; remove the remaining voltage source and replace it with its internal resistance of zero ohms as shown in Figure 10-2.
2. Next draw current arrows that show current flow based on the source polarity as shown in Figure 10-3.
3. Calculate the current that flows through each resistor and label the current arrows. See Figure 10-4.
   (a) Since the 3 and 6 ohm resistors are in parallel, the total circuit resistance is 14 ohms. Therefore, the total current is 1 ampere.
   (b) Using the current divider equation, the individual currents are:

$$I_{3\,\Omega} = \frac{1\,A \times 6\,\Omega}{9\,\Omega} = 0.667\,A \qquad\qquad I_{6\,\Omega} = \frac{1\,A \times 3\,\Omega}{9\,\Omega} = 0.333\,A$$

CLEAR PLASTIC SHEET #1

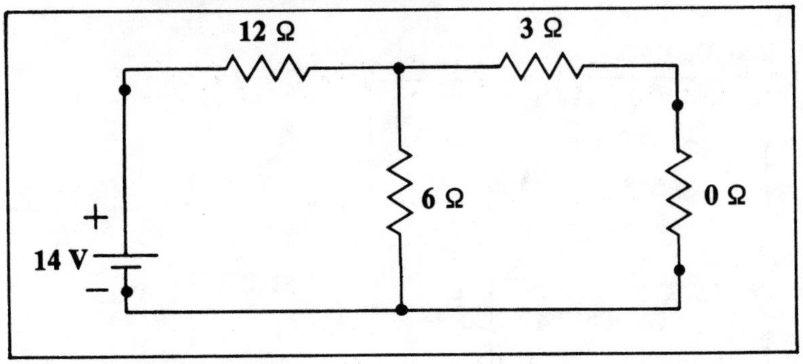

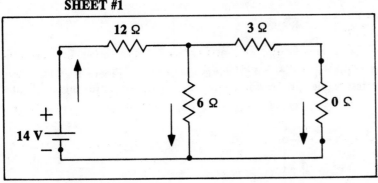

FIGURE 10-2

SHEET #1

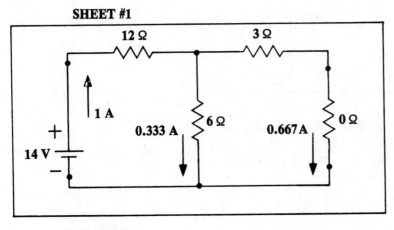

FIGURE 10-3

SHEET #1

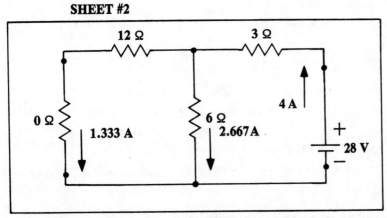

FIGURE 10-4

SHEET #2

FIGURE 10-5

4. Repeat Steps 1 through 3 for the remaining voltage source. Figure 10-5 shows the result of applying Steps 1 through 3 to the second source. Remember to use one clear plastic sheet for each modified circuit.

5. After all modified circuits have been completed, place all plastic sheets together and observe the current arrows as noted in Figure 10-6.

SHEET #1 PLACED OVER SHEET #2

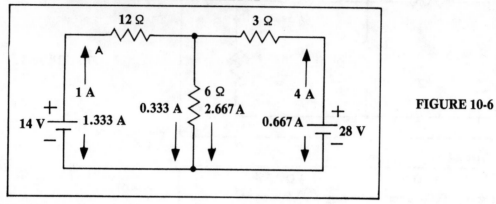

FIGURE 10-6

6. The algebraic sum of the currents of the superimposed drawings are shown in Figure 10-7. Observe that the current arrow having the greatest value predominates in the composite drawing of Figure 10-7.

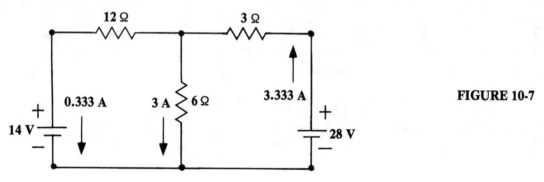

FIGURE 10-7

ANALYSIS OF A CIRCUIT CONTAINING THREE SOURCES

The conditions exhibited in Figure 10-8(a) would occur for a 3-source network. Currents B and C are added together since they are both going in the same direction, see Figure 10-8(b), and the resulting current is the algebraic sum of I_A and $I_B + I_C$.

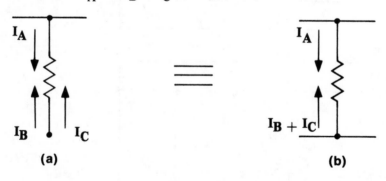

(a)   (b)

FIGURE 10-8

SUGGESTED OBSERVATIONS

1. Check each circuit to see that no wiring errors have been made.
2. Take time to make circuit hook-ups neat and easy to follow.
3. Record all readings and calculations in the data tables.

76

LIST OF MATERIALS

1. VOM
2. MFM
3. VPS
4. Four 1.5 volt Batteries
5. Resistors—all at least ½ watt

47 Ω	150 Ω	1.5 kΩ
56 Ω	330 Ω	2.2 kΩ
100 Ω	1 kΩ	

EXPERIMENTAL PROCEDURE

In this experiment, you will use the superposition theorem to compute the currents in several networks. You will then construct the circuits and measure the currents in the networks.

SECTION A: SUPERPOSITION THEOREM

1. Using the superposition theorem, **calculate** the current in each resistance of Figure 10-9.

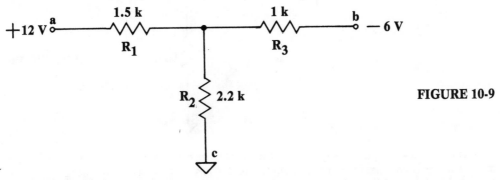

FIGURE 10-9

2. Reproduce the circuit of Figure 10-10 in the data tables; also draw in the appropriate current arrows. Observe that the 6 volt source of Figure 10-9 has been replaced with its internal resistance of zero ohms.

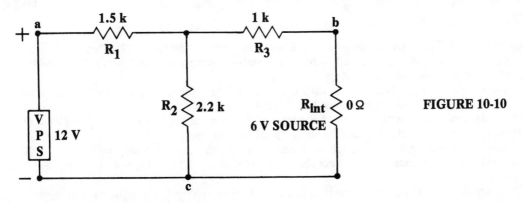

FIGURE 10-10

3. Construct the circuit of Figure 10-10.
4. Using a milliammeter, measure the current passing through each resistor. Label each of the current arrows in Step 2 with its measured current value.
5. Using Figure 10-11, complete the superposition theorem by repeating Steps 2 through 4.
6. Draw the circuit of Figure 10-9 in the data tables. Make a composite of the current arrows developed in Steps 2 through 5, using the currents of Figures 10-10 and 10-11.
7. Algebraically determine the resulting currents passing through each resistor of Step 6.
8. Using two voltage sources, construct the circuit of Figure 10-9 and measure the current passing through each resistance.
9. Using Equation 4-7, % error equation, compare the values of current obtained in Step 8 to that of Step 7.
10. Using the % error equation, compare the values of current of Step 8 to that of Step 1.

77

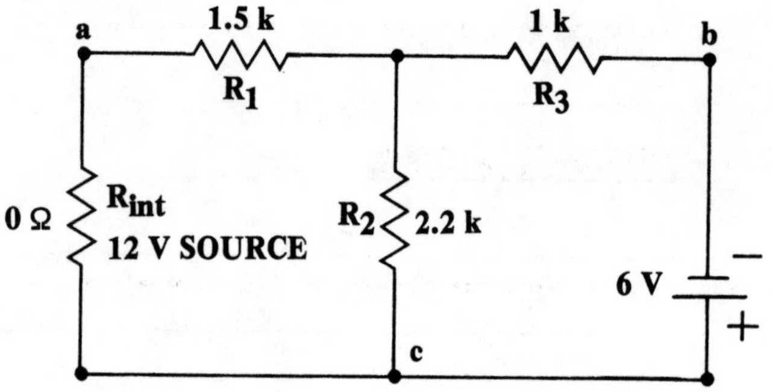

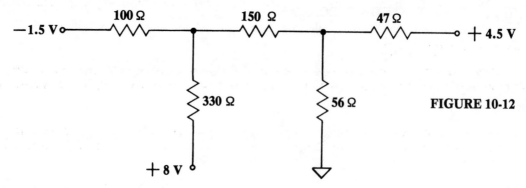

FIGURE 10-11

SECTION B: SUPERPOSITION THEOREM

1. Using the superposition theorem and the circuit of Figure 10-12, calculate the current in each resistance.

FIGURE 10-12

2. Redraw the circuit of Figure 10-12 in the data table, with the 4.5 and 8 volt source replaced with their internal resistance (zero ohms). Include the current arrows in this drawing.
3. Construct the circuit of Step 2. Using the MFM, measure the voltage across each resistor and then calculate the current in each resistance.

NOTE: This method of determining current is used to prevent the circuit from being loaded by a milliammeter.

4. Label each of the current arrows in Step 2 with the current determined in Step 3.
5. Redraw the circuit of Figure 10-12 with the 1.5 and 4.5 volt source replaced with their internal resistance of zero ohms and repeat Steps 2 through 4.
6. Redraw the circuit of Figure 10-12 with the 1.5 and 8 volt source replaced with their internal resistance of zero ohms and repeat Steps 2 through 4.
7. In the data tables, draw the circuit of Figure 10-12. Using the current arrows from each of the three drawings, and their values, make a composite of the currents.
8. Obtain the current passing through each resistor by algebraically adding the currents of Step 7.
9. Using three voltage sources, construct the circuit of Figure 10-12 and measure the current in each resistance.
10. Compare the values of current obtained in Step 9 to that in Step 8. Use the % error equation for this comparison.
11. Using the % error equation, compare the values of current in Step 9 to that in Step 1.

DATA INTERPRETATION AND CONCLUSIONS

Write a general summary of the ideas presented in this experiment and include your own conclusions.

APPLICATIONS

Although the superposition theorem can be used to solve multi-source networks, the main advantage will become apparent in networks containing both alternating and direct currents.

PROBLEM

1. Determine the power ratings of the resistors in Figure 10-1.

11 | THEVENIN'S THEOREM

OBJECTIVES

Thevenin's Theorem provides a method of analysing a circuit by reducing a complex circuit to a single source and a single resistance. This experiment explores the techniques of reducing the complex circuit to the simple Thevenin equivalent circuit. The theorem is verified experimentally and applications are investigated.

THEVENIN'S THEOREM

THEVENIN'S THEOREM states: **Any linear bilateral two-terminal network, regardless of complexity, can be replaced with an electrically equivalent circuit consisting of one source (E_{th}) and one impedance (R_{th}).** See Figure 11-1.

NOTE: The term impedance encompasses reactive devices which are encountered in alternating current circuits. Since no reactive devices are presented in direct current circuit analysis, the word impedance can be replaced by the word resistance.

Although Thevenin's treatment of a network is based on a mathematical analysis, under certain controlled conditions, the individual circuit parameters can be measured.

Examine the network of Figure 11-2. Notice that terminals **a** and **b** are connected to the 12 ohm resistor. The voltage across this resistor which is E_{th} can be determined with the voltage divider equation. Thus:

$$E_{th} = E_{12\ \Omega} = \frac{36\ V \times 12\ \Omega}{18\ \Omega} = 24\ V$$

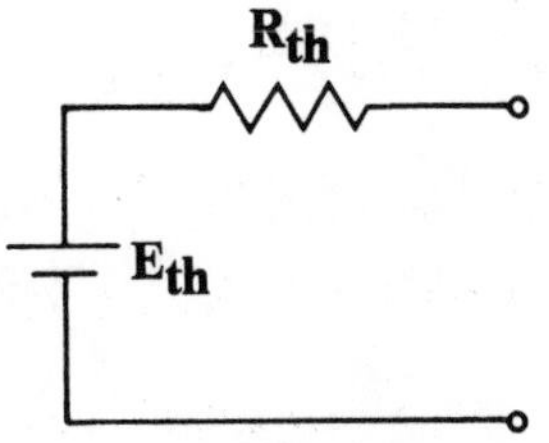

FIGURE 11-1

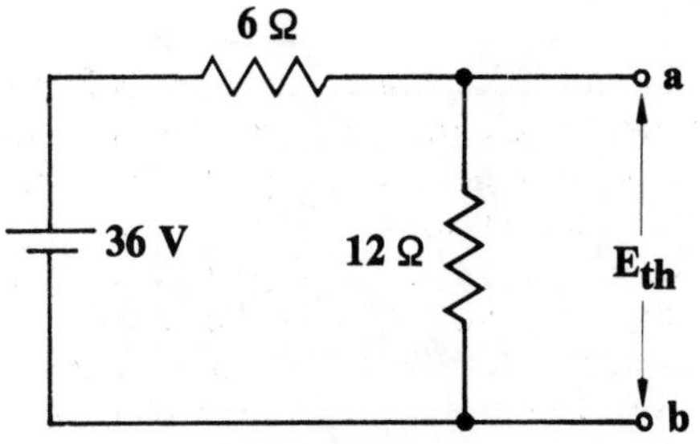

FIGURE 11-2

If a voltmeter, having a very high internal resistance so as to prevent loading of the circuit, were to be connected across terminals **a** and **b**, then the voltmeter would indicate an E_{th} of 24 volts.

This voltage is now transferred to Figure 11-3, the Thevenized equivalent of Figure 11-2. R_{th} can be determined by replacing the 36 V source with its internal resistance of 0 Ω as shown in Figure 11-4 and computing the resistance that would be seen looking back into the circuit from terminals **a** and **b**. The circuit has taken the form of a simple series-parallel configuration. Therefore, the internal resistance (R_{th}) can be determined from:

$$R_{th} = \frac{R_1 R_2}{R_1 + R_2} = \frac{12\ \Omega \times (6\ \Omega + 0\ \Omega)}{12\ \Omega + 6\ \Omega + 0\ \Omega} = \frac{72\ \Omega}{18\ \Omega} = 4\ \Omega$$

This resistance is now transferred to Figure 11-3.

The remaining quantity (not shown) is the short circuit current (I_{sc}). This value can be determined by either Ohm's law or controlled measurements.

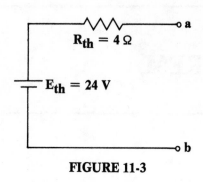

FIGURE 11-3

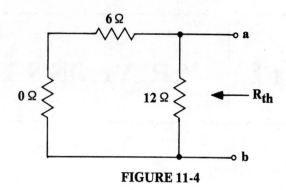

FIGURE 11-4

OHM'S LAW

Using the information given in Figure 11-3, it can be seen that the maximum amount of current that could flow is limited by the 4 ohm resistor when terminals **a** and **b** are connected together with a load of zero ohms resistance. See Figure 11-5. Thus, the short circuit current is:

$$I_{sc} = 24 \text{ V} / 4 \,\Omega = 6 \text{ A}$$

If a load of zero ohms is connected across terminals **a** and **b** of Figure 11-6, then the maximum current that could flow is:

$$I_{sc} = 36 \text{ V} / 6 \,\Omega = 6 \text{ A}$$

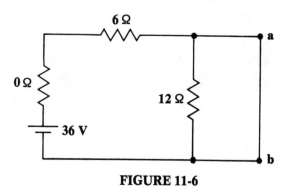

FIGURE 11-5

FIGURE 11-6

CONTROLLED MEASUREMENT

If an ammeter, with an internal resistance of zero ohms, is connected across terminals **a** and **b**, the same results of 6 A would be obtained. Because an ammeter with zero resistance and a load with zero resistance are the same condition, the current in the circuit is the same.

THEVENIZING A CIRCUIT

The best use of the Thevenin's equivalent circuit concept results when you begin to visualize a complex circuit as several simple circuits, as shown in Figure 11-7. The following example demonstrates the use of Thevenin's theorem.

EXAMPLE 1

Determine the current through the 10 ohm resistor in Figure 11-7.

SOLUTION: Observe in Figure 11-8 that by placing x's to the right hand side of the 4 ohm resistor two separate circuits are created. Compare Figure 11-8 with Figure 11-2 and note their similarity. Thus, the left hand portion of Figure 11-8 is replaced with a Thevenized equivalent as shown in Figure 11-9. After electrically rejoining the two separate circuits as noted in Figure 11-10, you may see that the circuit of Figure 11-7 has been reduced to a simple series circuit. See Figure 11-11. Observe how each subsequent reduction has brought the circuit closer to the **simple** series circuit of Figure 11-12. Figure 11-12 results when Kirchhoff's closed loop concept is applied to Figure 11-11. With the values of 4.67 volts and 12.67 ohms, the current will be 369 mA. Since the 10 ohm resistance is part of the 12.66 ohms in Figure 11-11,

80

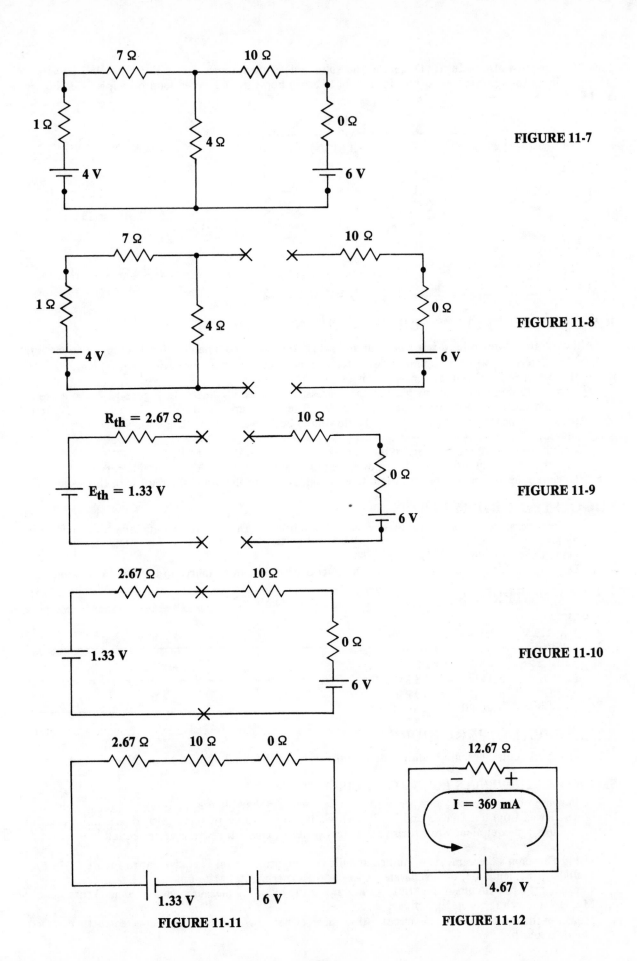

7 Ω   10 Ω

1 Ω

4 Ω   0 Ω

4 V   6 V

FIGURE 11-7

7 Ω   10 Ω

1 Ω

4 Ω   0 Ω

4 V   6 V

FIGURE 11-8

R_{th} = 2.67 Ω   10 Ω

E_{th} = 1.33 V   0 Ω

6 V

FIGURE 11-9

2.67 Ω   10 Ω

1.33 V   0 Ω

6 V

FIGURE 11-10

2.67 Ω   10 Ω   0 Ω

1.33 V   6 V

FIGURE 11-11

12.67 Ω

I = 369 mA

4.67 V

FIGURE 11-12

81

then 369 mA must also be in it. Therefore, the voltage developed across the 10 Ω resistor is 3.69 volts. The remaining voltage drops are determined by once again applying the closed loop principle. See Figure 11-13.

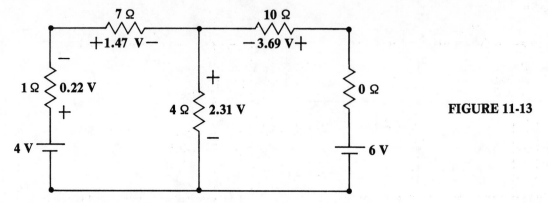

FIGURE 11-13

RECOMMENDED STEPS FOR THEVENIZING A CIRCUIT

Although there are no **hard-fast** rules, the following steps are recommended for Thevenizing a circuit:

1. First determine which components are directly concerned with the unknown quantity, e.g., current through a resistor, voltage across a resistor., etc.
2. Remove those components by placing an x across the connecting lines. In Figure 11-7, the current through the 10 Ω resistor was desired; however, the 6 V source is also in series with the 10 ohm resistor. Since the same current must be flowing in both components, then both components are removed.
3. Reduce the remaining circuitry to the Thevenin's equivalent of Figure 11-1.
4. Join the equivalent circuit to the removed circuit and solve for the desired circuit parameter.
5. Once the desired parameter has been determined, **return** to the original circuit with this value.

SUGGESTED OBSERVATIONS

1. Make several drawings to express new circuit conditions. Do not crowd individual drawings with data.
2. Do not short circuit the VPS.
3. Do not attempt to make resistance measurements with the power source in the "on" condition.

LIST OF MATERIALS

Alternate Materials:

1. MFM
2. Two 1.5 V Batteries
3. VPS
4. Resistors—all at least ½ watt
   1.8 kΩ   3.3 kΩ   6.8 kΩ
   2.2 kΩ   4.7 kΩ   10 kΩ
5. Test Circuit: Consult Your Instructor

EXPERIMENTAL PROCEDURE

In this experiment, Thevenin's equivalent circuit will be verified and applied to the **black-box** concept.

SECTION A: THEVENIN'S EQUIVALENT CIRCUIT

The purpose of this part of the experiment is to demonstrate that Figures 11-14(a) and (b) are equivalent when viewed from terminals **a** and **b**. Thevenin's theorem is used to accomplish this. As previously stated, circuits are equivalent when their parameters (voltage, current, resistance) are equal.

1. Using Thevenin's theorem, show mathematically that Figure 11-14(a) is equivalent to Figure 11-14(b). Include this calculation, on a separate paper, in the experimental write-up.
2. Construct the two circuits of Figure 11-14 and make the following measurements in Steps 3 through 5.
3. Measure the voltage drop E_{ba} of each circuit.
4. Measure the current, of each circuit, that flows when an ammeter is connected between terminals **a** and **b**.

5. Replace each source with its internal resistance. Since the internal resistance is 0 Ω, a wire is used for this connection. Measure the resistance across terminals **a** and **b**.
6. Using Equation 4-7, % error equation, compare the parameters of Figure 11-14(a) to those of Figure 11-14(b).

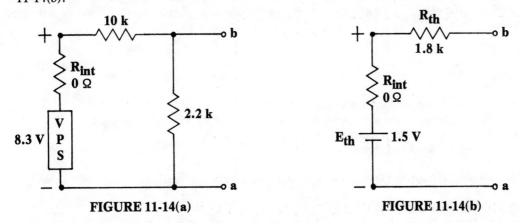

FIGURE 11-14(a)   FIGURE 11-14(b)

SECTION B: THEVENIN'S THEOREM VERIFICATION

1. Using Thevenin's theorem, mathematically determine the voltage (E_0) across the load resistance.
2. Construct the circuit of Figure 11-15 to verify the computations of Step 1. Measure E_{th} by removing R_L and then place the voltmeter between points **a** and **b**. Then replace the load and measure E_0.

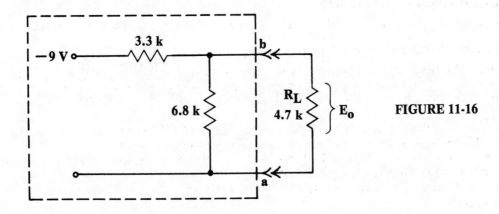

FIGURE 11-15

3. Draw and label the Thevenin's equivalent circuit of Figure 11-15. Construct this circuit and measure E_0.
4. Using the % error equation, compare the value of E_0 obtained in Step 1 to that obtained in Step 3.

SECTION C: APPLICATION OF THEVENIN'S THEOREM

Many circuits are sealed in an enclosure and only the output terminals are available for measurements. It is not possible to Thevenize these circuits with pencil and paper. Instead, the following procedure is used.

1. Construct the circuit of Figure 11-16 without the load R_L connected.

FIGURE 11-16

2. Assume that the circuit is inaccessible and only the terminals are available for measurement. Begin to reduce the circuit to an equivalent by measuring the open circuit terminal voltage (E_{ba}). The open circuit voltage is E_{th} ($E_{ba} = E_{th}$). Place R_L across terminals **a** and **b** and measure E_o.
3. Determine R_{th} by applying Equation 11-1. Since the circuit is enclosed, it is impossible to measure R_{th} by physically replacing the source with its internal resistance; therefore, R_{th} must be calculated using Equation 11-1.

$$R_{th} = \left[\frac{E_{th}}{E_o} - 1 \right] R_L \qquad \qquad \textbf{EQUATION 11-1}$$

where: R_{th} = the equivalent resistance
E_{th} = the open circuit voltage measured across the unloaded terminals
E_o = the voltage measured across the load
R_L = the load resistance

4. Draw the equivalent circuit of Figure 11-16. Label each component with its determined value.

SECTION D: BLACK BOX CONCEPT

1. Determine the Thevenin's equivalent circuit of the **black box test circuit** (supplied by your instructor) by taking measurements at the output terminals. Since it was impractical to build the voltage source into the black box, 9 volts must be supplied to the terminals marked input. The load is 10 kΩ.
2. Draw the equivalent circuit in the space provided in the data table. Clearly label each component, including the load resistance, with its value.
3. Using the schematic of Step 2, calculate the power dissipated by the load resistance.

DATA INTERPRETATION AND CONCLUSIONS

Write a general summary of the ideas presented in this experiment. This discussion should include:
1. A comparison between the original circuit and the Thevenin's electrical equivalent.
2. The procedure required to reduce a **black box** to Thevenin's equivalent.
3. Your own conclusions.

APPLICATIONS

Thevenin's equivalent circuit is used in circuit analysis as well as electrical description (modeling) of: (1) vacuum tubes, (2) semiconductors, and (3) signal generators.

PROBLEMS

1. Thevenize the circuit of Figure 11-17 from Point **a** to the reference point.

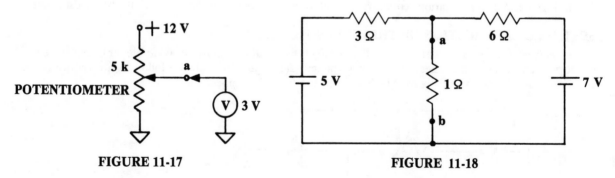

FIGURE 11-17   FIGURE 11-18

2. In your own words, describe the step-by-step procedure needed to Thevenize the circuit of Figure 11-18 between **a** and **b**.

12 ‖ MAXIMUM POWER TRANSFER

OBJECTIVES

The Maximum Power Transfer Theorem states the circuit conditions for dissipating the greatest amount of power in the load from a source. This experiment graphically demonstrates maximum power transfer and shows the need for the transfer of maximum power.

THE MAXIMUM POWER TRANSFER THEOREM

THE MAXIMUM POWER TRANSFER THEOREM states: **Maximum power is transferred from the generator (source) to the load when the load impedance (resistance) is equal to the internal impedance (resistance) of the generator.** This concept is illustrated with the aid of Figure 12-1.

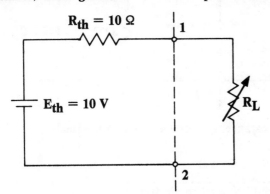

FIGURE 12-1

Observe that the portion of the circuit to the left of terminals 1 and 2 is in the form of a Thevenin's equivalent circuit—one source in series with one resistance. To prove that maximum power will be dissipated by R_L when its value equals the 10 ohms of the source, express the power dissipation as:

$$P_L = I^2 R_L \qquad \text{and} \qquad I = \frac{E_{th}}{R_{th} + R_L} \qquad \text{Thus:}$$

$$P_L = \left(\frac{E_{th}}{R_{th} + R_L} \right)^2 R_L = \frac{E_{th}^2 R_L}{R_{th}^2 + 2R_{th}R_L + R_L^2}$$

Thus, for the circuit of Figure 12-1, where $E_{th} = 10$ V and $R_{th} = 10$ Ω, $E_{th}^2 = 100$ and $R_{th}^2 = 100$:

$$P_L = \frac{100R_L}{100 + 20R_L + R_L^2}$$

Table 12-1 shows the values of P_L for five values of R_L. Graphing this data shows that a peak or maximum point occurs when $R_L = 10$ ohms. See Figure 12-2. The graph shows that maximum power was being dissipated when $R_L = R_{th}$.

MATCHING LOAD AND SOURCE RESISTANCE

There are several circuits that will not function properly unless the load impedance is matched to the source resistance. When such a circuit is encountered, every precaution must be taken to assure that a

TABLE 12-1

R_L	P_L
8 Ω	2.47 W
9 Ω	2.49 W
10 Ω	2.50 W
11 Ω	2.49 W
12 Ω	2.47 W

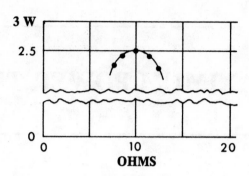

FIGURE 12-2

proper match is maintained at all times.

For example, a certain piece of equipment is in the shop for a routine calibration check. One of the tests consists of coupling a 1 volt signal into a 50 ohm input terminal. However, the shop generator has a fixed level output of 10 volts at 50 ohms. A special circuit is required so that both the generator and the equipment under test can "see" a 50 ohm load. In addition, the input signal level to the equipment under test must be reduced to 1 volt. Such a circuit is depicted in Figure 12-2.

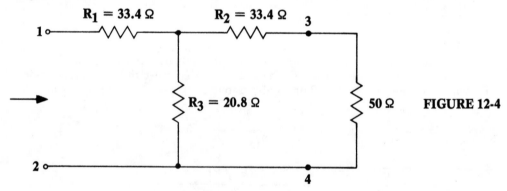

FIGURE 12-3

The values of R_1, R_2, and R_3 must be such that the resistance, as seen looking into terminals 1 and 2 will be 50 ohms. See Figure 12-4.

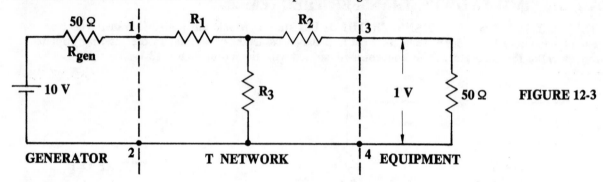

FIGURE 12-4

Using the values shown, the resistance that would be measured between terminals 1 and 2 is:

$$R_1 + \frac{R_3 \ (R_2 + 50)}{R_3 + R_2 + 50} = 33.4 \ \Omega + \frac{20.8 \ \Omega (33.4 \ \Omega + \ 50 \ \Omega)}{20.8 \ \Omega + \ 33.4 \ \Omega + 50 \ \Omega} = 33.4 \ \Omega + \ 16.6 \ \Omega = 50 \ \Omega$$

Therefore, the generator "sees" a 50 ohm load. If the resistance is measured across terminals 3 and 4 in Figure 12-5, the resistance will be

$$R_2 + \frac{R_3(R_1 + 50)}{R_3 + R_1 + 50}$$

Since R_1 is equal to R_2, this is the same equation that was used to determine the resistance between terminals 1 and 2. Thus, the load also "sees" a 50 ohm load.

86

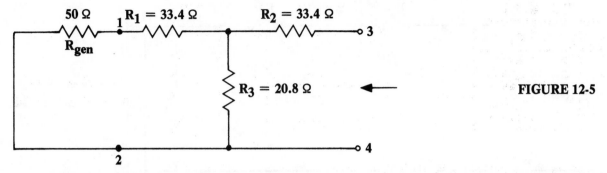

FIGURE 12-5

In addition to maintaining a match between the generator and the load, the generator signal had to be reduced (attenuated) to one volt. Thevenizing the circuit of Figure 12-6 as shown, results in Figure 12-7 with a 1 volt signal appearing across the 50 ohms.

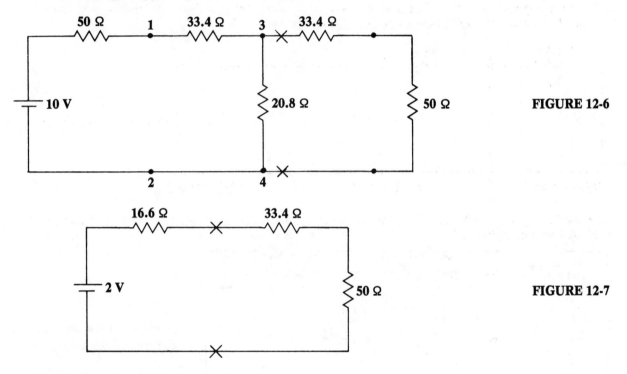

FIGURE 12-6

FIGURE 12-7

The voltage across the 50 ohm load is:

$$E_{R_L} = \frac{2 \times 50}{100} = 1 \text{ V}$$

T NETWORKS FOR MATCHED GENERATOR AND LOAD RESISTANCE

The actual procedure used to determine resistance values for T networks will be covered in a more advanced course. However, a T network that will maintain a match between identical generator and load resistance can be discussed at this time. For the conditions just stated, as shown in Figure 12-8, R_1 is always equal to R_2. Observe that the circuitry to the right of terminals 1 and 2 can be replaced by a 50 ohm resistor. See Figure 12-9. Obviously the voltage developed across either 50 ohm resistance is 5 volts and the current is 100 mA. Also the current through the 50 ohm load to the right of terminals 3 and 4 of Figure 12-8 is 20 mA. These currents and a voltage loop are shown in Figure 12-10. Starting at terminal 2 and moving in the clockwise direction, the loop equation is:

$$10 - 0.1(50) - 0.1R_1 - 0.02R_1 - 0.02(50) = 0$$

NOTE: $R_2 = R_1$.

87

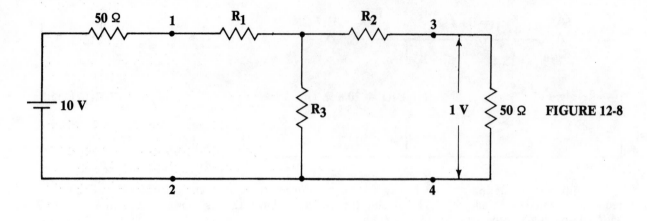

FIGURE 12-8

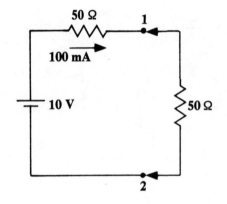

FIGURE 12-9

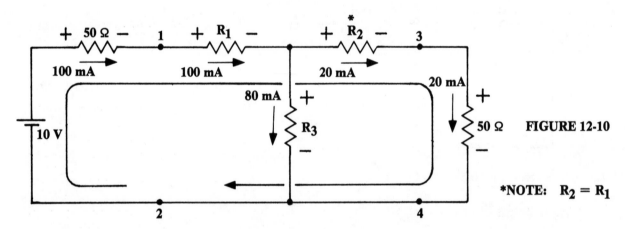

FIGURE 12-10

*NOTE: $R_2 = R_1$

Simplifying and solving for R_1 results in the following:

$$10 - 5 - 1 = 0.12R_1 \qquad \text{and} \qquad R_1 = 4/0.12 = 33.3 \ \Omega$$

The voltage across R_3 is 1.667 V when the current is 80 mA and R_3 is:

$$R_3 = 1.667 \ \text{V}/80 \ \text{mA} = 20.8 \ \Omega$$

NOTE: These are the values that were used in Figure 12-6.

SUGGESTED OBSERVATIONS
1. Make all calculations, schematics, and graphs as neat as possible.
2. Record all measurements and calculations in the data tables.

LIST OF MATERIALS

1. Photovoltaic Cell (Solar Cell)
2. 75 to 100 watt Light Source
3. MFM
4. Resistors—all at least ½ watt

100 Ω	3.3 kΩ	15 kΩ
1 kΩ	5.6 kΩ	27 kΩ
1.2 kΩ (two)	6.8 kΩ	39 kΩ
2.2 kΩ	10 kΩ (two)	
2.7 kΩ	12 kΩ	

5. VPS
6. Graph Paper (10 × 10 to the inch)

Alternate Materials

_____

_____

_____

_____

_____

EXPERIMENTAL PROCEDURE

This experiment provides empirical verification of the maximum power transfer theorem. The maximum power transfer characteristic of a photovoltaic cell is determined. A T network is used to provide both maximum power transfer (impedance matching) and signal reduction (attenuation).

SECTION A: MAXIMUM POWER TRANSFER—GRAPHIC PROOF

This section of the experiment will demonstrate that the power developed in the load is at a maximum at only one particular value of load resistance.

1. Construct the circuit of Figure 12-11.

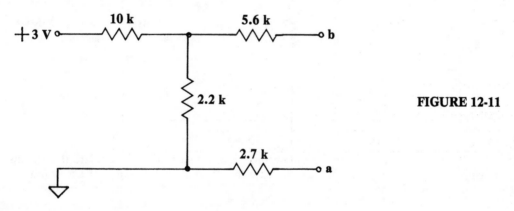

FIGURE 12-11

2. Measure the voltage drop, E_{ba}, for each of the following loads connected between terminals **a** and **b**:

   1 kΩ, 3.3 kΩ, 6.8 kΩ, 10 kΩ, 12 kΩ, 15 kΩ, 27 kΩ, 39 kΩ

3. Using Equation 9-2(b), calculate the power dissipated by each load.
4. Make a table of values for load resistance versus power dissipation.
5. Using the table of Step 4, plot a graph of power versus resistance. The graph and table of values should be included in the experimental write-up.
6. Using the graph, determine which value of resistance has maximum power transfer.

SECTION B: MAXIMUM POWER TRANSFER BY THEVENIN'S THEOREM

When the equivalent resistance, R_{th}, is equal to the load resistance, then the condition for maximum power transfer is met.

1. Construct the circuit of Figure 12-11 and experimentally determine the Thevenin's equivalent circuit using the method shown in Experiment 11.
2. Draw a schematic of the Thevenized circuit.
3. Using the % difference equation, compare the value of R_{th} determined in Step 1 to the value of resistance determined in Step 6 of Section A.

SECTION C: DETERMINING MAXIMUM POWER TRANSFER—SOLAR CELL

1. Place a solar cell 3 to 4 inches from a 75 to 100 watt lamp.

2. Measure the loaded and unloaded voltage output of the solar cell.

3. Determine the internal resistance of the solar cell by substituting into Equation 4-6. Select 100 Ω for R_L.

4. Using several values of load resistance centered around the value of the internal resistance determined in Step 2, set up a table of values and plot a curve for power versus resistance (as in Section A). Include the table of values and the curve in the experimental write-up.

5. Using the % difference equation, compare the value of load resistance for maximum power transfer determined in Step 2 to that value determined from the graph of Step 3.

6. Draw a schematic of the test set-up. Include the symbol for the solar cell, internal resistance, and load resistance.

SECTION D: ATTENUATOR PROBLEM

Design a T network which will give a 10 to 1 voltage attenuation with maximum power transfer. Before constructing the circuit of Figure 12-12, calculate the value of R_1, R_2, and R_3. Since maximum power is to be transferred, the load must work into 1.2 k ohms while the voltage source with its internal resistance must also see 1.2 k ohms.

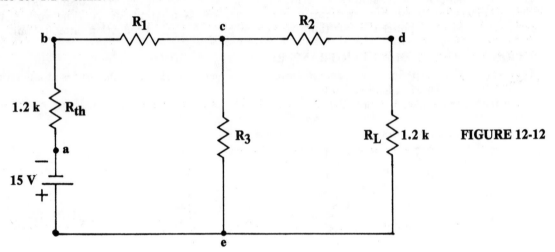

FIGURE 12-12

1. Draw the circuit of Figure 12-12 in the data table. Using Figure 12-13, calculate the amount of current flowing out of the voltage source. To indicate this current, place a current arrow labeled with its value on the drawing in the data table.

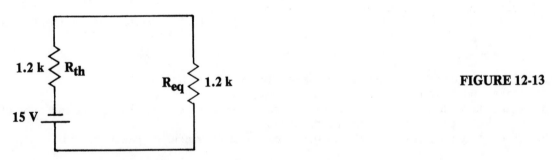

FIGURE 12-13

2. Determine the amount of current in the load. Since this is a 10 to 1 attenuator, the output voltage, E_{ed} in Figure 12-12, must be 0.1 of the source voltage. Place this current arrow on the figure in the data table.

3. Determine the current passing through R_2 of Figure 12-12. Enter this current arrow on the drawing in the data table.

4. Since this is a symmetrical T network, R_1 is equal to R_2. (Figure 12-12). Write a loop equation and solve for R_1 (let $R_2 = R_1$). Starting at Point **a** and moving clockwise (CW):

$$E_{ba} + E_{cb} + E_{dc} + E_{ed} - E_{ae} = 0$$

90

5. Solve for R_3.
6. Construct the circuit of Figure 12-12. Use the calculated values for the T network. (A pot may be used for R_3 while R_1 may be built by paralleling two values of resistance.)
7. Determine by measurement that the circuit design requirements have been met. That is, R_{th} is 1.2 kΩ when viewed from terminals **d** and **e** and the voltage drop across the load is 1.5 V.

DATA INTERPRETATION AND CONCLUSIONS

Write a general summary of the ideas presented in this experiment. This discussion should include:
1. The conditions for maximum power transfer.
2. The problems involved with attenuating a voltage source while maintaining maximum power transfer.
3. Your own conclusions.

APPLICATIONS

Some of the applications where maximum power must be transferred are as follows: (1) the antenna input terminals on the TV set, the hookup wire between the TV set and the antenna, and the antenna must all have the same impedance characteristics, (2) amateur and CB radio transmitters—the feedline and the antenna must all be matched, (3) microphone impedance matched to the input impedance of tape recorders and public address systems.

PROBLEMS

1. In the following circuit, determine the value of load resistance needed for maximum power transfer.

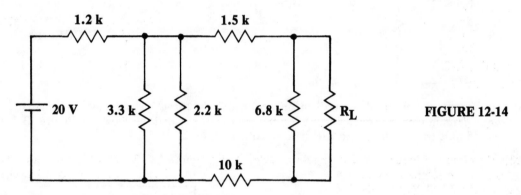

FIGURE 12-14

2. Determine the power dissipated by the load of problem 1.

91

13 | NORTON'S THEOREM

OBJECTIVES

Norton's Theorem provides a method of reducing a complex circuit to a constant current source and a conductance for the purpose of circuit analysis. This experiment validates Norton's Theorem. Constant current and constant voltage generators are also investigated.

NORTON'S THEOREM

NORTON'S THEOREM states: **Any linear, bilateral, two-terminal network can be reduced to a constant current generator in parallel with an equivalent resistance.** Norton's equivalent, however, unlike the Thevenin equivalent, cannot be physically constructed in the form of its symbolic schematic. See Figure 13-1.

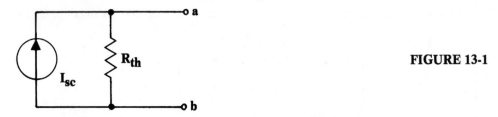

FIGURE 13-1

Maximum benefits are obtained from a Norton's equivalent by using it strictly as a mathematical tool. The fact that Thevenin's and Norton's circuits can be electrically equated, as shown in Figure 13-2, results in a relatively simple transfer from one circuit to the other.

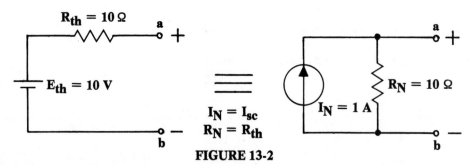

FIGURE 13-2

If the two circuits are electrically equivalent, then the polarities of the terminals **a** and **b** must also be identical. Thus, observe how the current arrow shown within the generator symbol indicates that current flow is from negative to positive within the source and positive to negative external to the source. Also, note that the maximum current that the Thevenized circuit of Figure 13-2 can deliver is 1 ampere and would occur when terminals **a** and **b** are connected together. This current is referred to as the **short circuit current** (I_{sc}). This is the same current that would flow in Norton's equivalent circuit when terminals **a** and **b** are connected together. The usefulness of Norton's equivalent and the relative ease of converting from Thevenin's to Norton's circuits is shown in Example 1.

EXAMPLE 1:

Determine the current flowing in the 2 ohm resistor of Figure 13-3.

SOLUTION: Section the network as shown in Figure 13-4. The two Thevenin's equivalent circuits, No.'s 1 and 2, can now be converted to Norton's equivalent circuits as shown in Figure 13-5. The short

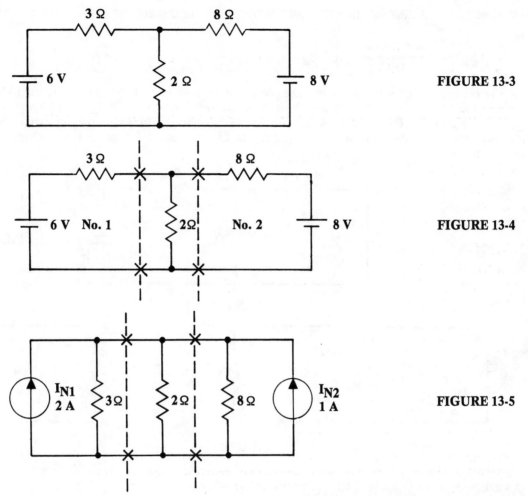

FIGURE 13-3

FIGURE 13-4

FIGURE 13-5

circuit currents are:

$$I_{N_1} = I_{sc_1} = \frac{6\ V}{3\ \Omega} = 2\ A \qquad \text{and} \qquad I_{N_2} = I_{sc_2} = \frac{8\ V}{8\ \Omega} = 1\ A$$

The network can be further rearranged as shown in Figure 13-6 for ease in obtaining the total current of the two current generators.

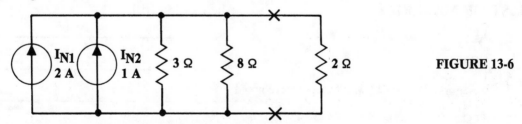

FIGURE 13-6

The algebraic sum of these currents is 3 amperes. See Figure 13-7. The 2.18 ohms is the equivalent resistance of the parallel combination of the 3 and 8 ohm resistors.

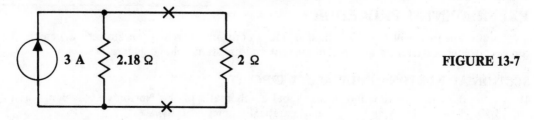

FIGURE 13-7

93

Using the current divider equation (Equation 5-7), the current through the 2 ohm resistor is

$$I_{2\,\Omega} = \frac{3\,A \times 2.18\,\Omega}{4.18\,\Omega} = 1.565\,A$$

If the polarity of the 8 V battery in Example 1 had been reversed as shown in Figure 13-8, then the current generator arrow would also have been reversed and the network of Figure 13-6 would then have appeared as shown in Figure 13-9. Again, adding the currents algebraically, the net current is 1 ampere as shown, and the final current arrow is indicating the direction of the 2 A generator which has the highest magnitude.

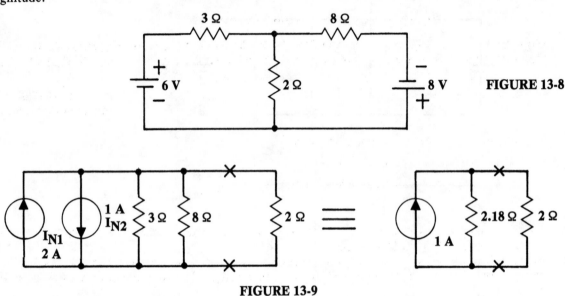

FIGURE 13-8

FIGURE 13-9

SUGGESTED OBSERVATIONS

1. Thoroughly read through each experiment section prior to working the experiment.
2. Check the meter range and function setting prior to making any measurements.
3. **Do not attempt to short circuit** either the power supplies or batteries.
4. Remove the power from the circuit prior to working on it.
5. Record all data in the data tables.

LIST OF MATERIALS   Alternate Materials

1. VPS
2. MFM
3. VOM
4. 1.5 V Cell
5. Resistors—at least ½ watt except where noted

10 Ω	330 Ω	2.7 kΩ
47 Ω	470 Ω	4.7 kΩ
100 Ω	1 kΩ (two - 1 W)	10 kΩ (2 W
220 Ω	1.8 kΩ (1 W)	1 MΩ

EXPERIMENTAL PROCEDURE

Norton's Theorem will now be validated and a constant current generator constructed. You will also compare the characteristics of constant current and constant voltage generators.

SECTION A: NORTON'S EQUIVALENT CIRCUIT

1. In the data table, draw and clearly label a schematic of the Norton's equivalent circuit of Figure 13-10. Use the 4 mA as I_{sc} and mathematically determine R_{th}.

94

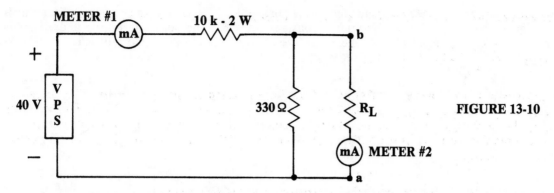

FIGURE 13-10

2. Using the Norton's equivalent circuit of Step 1 and the following equation (Equation 13-1), mathematically determine the load current for each of the following load values: 100 Ω, 220Ω, 470 Ω, 1 kΩ.

$$I_L = \frac{I_{sc}R_{th}}{R_{th} + R_L}$$

EQUATION 13-1
(Based on Equation 5-7)

where: I_L = current through the load

I_{sc} = short circuit current

R_{th} = parallel equivalent resistance

R_L = load resistance

3. Construct the circuit of Figure 13-10. With terminals **a** and **b** shorted, adjust the VPS for a milliammeter deflection of 4 mA. (This is I_N).

4. Remove the short of Step 3 and measure the current that passes through each of the four load resistors listed in Step 2 when each is connected to terminals **a** and **b**. Use two milliameters as noted in Figure 13-10.

5. Using the % error equation, compare each of the currents of Step 2 to those of Step 4.

6. During Step 4, did milliammeter one vary from the 4 mA value of I_N? If so, how much?

SECTION B: CONSTANT CURRENT CONCEPT

In Step 6 of Section A, the current variation was insignificant, and the circuit of Figure 13-10 was a constant current source. When the source internal resistance is at least 20 times greater than the load resistance then, the source is constant current.

1. With the milliammeter set to measure 1 mA, construct the circuit of Figure 13-11.

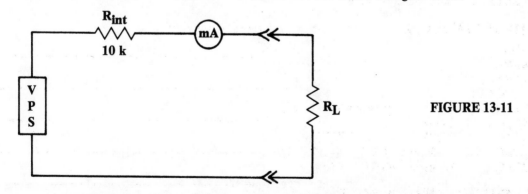

FIGURE 13-11

2. With the load terminals shorted, adjust the VPS for a milliammeter deflection of 1 mA.

3. Remove the short and record the milliammeter reading for each of the following values of load resistance: 10 Ω, 47 Ω, 100 Ω.

4. Is Figure 13-11 a constant current source?

SECTION C: CONSTANT CURRENT SOURCE SIMULATION

In Section B it was seen that under certain circuit conditions, constant current could be achieved. However, if constant current is to be maintained for a wide range of load values, then a circuit with flexibility is needed. Figure 13-12 is capable of delivering a constant current to the output terminals under varying load conditions, while Figure 13-13 is the electrical block diagram of a constant current source.

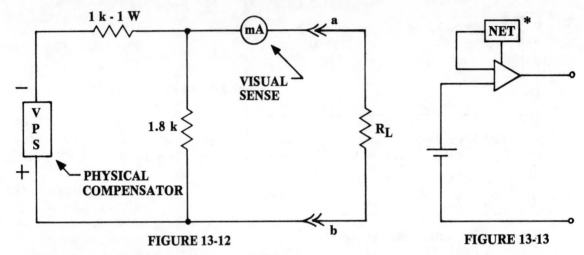

FIGURE 13-12   FIGURE 13-13

*Net is the abbreviation for Network and in Figure 13-13, it represents the feedback network required to compensate for current change.

1. Construct the circuit of Figure 13-12. With the output terminals **a - b** short circuited, adjust the VPS (physical compensator) for 3 mA indication on the milliammeter (visual sense).
2. Remove the short. Place a 47 ohm resistor (R_L) in the circuit. Record the reading of the visual sense unit. Then adjust the current to 3 mA.
3. Replace the 47 ohm resistor, in turn with each of the following values and repeat Step 2.

       100 Ω,   470 Ω,   1 kΩ,   2.7 kΩ.   4.7 kΩ,   10 kΩ

NOTE: This circuit simulates an electronic circuit which maintains a constant current for varying loads. In the electronic circuit, the sense and compensation are carried out automatically.

SECTION D: CONSTANT VOLTAGE

Constant voltage is achieved when the ratio of the source internal resistance to the load resistance is at least 1 to 10. Since the internal resistance of a dry cell is very low, the cell will exhibit constant voltage characteristics over a wide range of load resistances.

1. Construct the circuit of Figure 13-14.

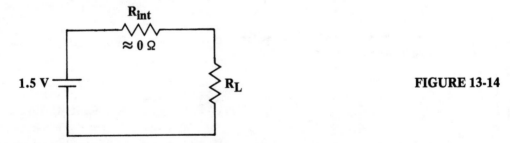

FIGURE 13-14

2. Record the voltage drop across the load for each of the following load resistances:

      470 Ω,   1 kΩ,   10 kΩ,   1 MΩ
3. Is the source of Figure 13-14 a constant voltage source?

DATA INTERPRETATION AND CONCLUSIONS

Write a general summary of the ideas presented in this experiment. This discussion should include:
1. Constant current generators.
2. Constant voltage generators.
3. The validity of Norton's theorem.
4. Your own conclusions.

APPLICATIONS

The following circuits show some applications for both Thevenin's and Norton's equivalent circuits. (1) Figure 13-15 shows the low frequency equivalent circuit of a triode vacuum tube. Observe that the section to the right of the dashed line is Thevenin's equivalent. (2) Figure 13-16 shows the low frequency equivalent circuit of a pentode vacuum tube. Observe that the section to the right of the dashed line is Norton's equivalent.

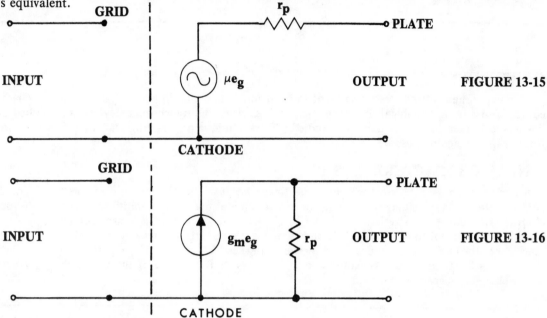

FIGURE 13-15

FIGURE 13-16

(3) Figure 13-17 shows the low frequency equivalent circuit of a transistor in the hybrid configuration. The circuitry to the left of the dashed line is Thevenin's equivalent and to the right is Norton's equivalent.

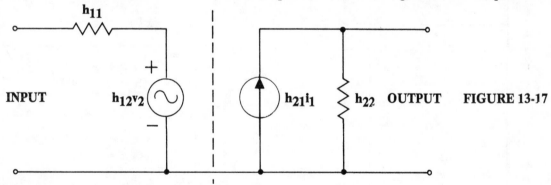

FIGURE 13-17

PROBLEMS

1. Using Norton's theorem, determine the current through the 8 ohm resistor of Figure 13.3.
2. What is the current through the 2 ohm resistor of Figure 13-9?
3. Determine whether the use of Thevenin's or Norton's theorem would be the easiest in obtaining a solution for the voltage E_{ag} of Figure 13-18.

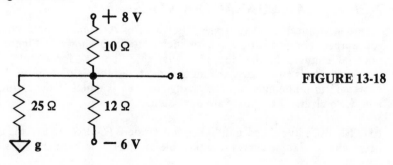

FIGURE 13-18

97

14 | NON-LINEAR DEVICES

OBJECTIVES

This experiment introduces the student to both linear and non-linear volt-ampere (E-I) characteristic curves for use in the graphical analysis of electronic devices. Four electronic devices are graphed and their linear and bilateral characteristics are studied. The technique of graphing and the art of extracting data from graphs is again stressed.

LINEAR CHARACTERISTICS

A device, such as a resistor, is said to be linear because the ohmic value of the resistance does not change with a change either in electrical or environmental operating conditions. The volt-ampere (E-I) characteristic curve for a resistor is shown in Figure 14-1. The plot is a straight line showing current directly proportional to the voltage. It shows that a change in ΔI_1 produced by the change in ΔE_1 is identical to the change in ΔI_2 produced by the change in ΔE_2 when ΔE_1 equals ΔE_2. In other words, the curve is linear and has a constant slope. It shows that the value of the resistance does not change under changed operating conditions.

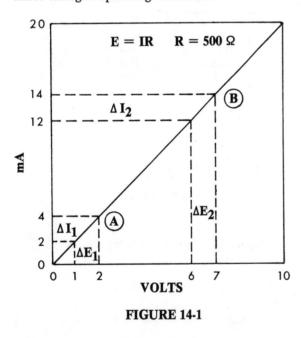

FIGURE 14-1

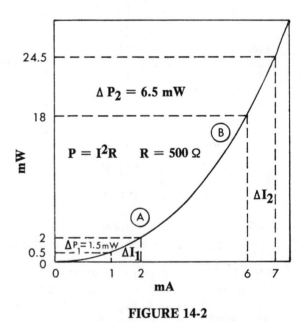

FIGURE 14-2

NON-LINEAR CHARACTERISTICS

In a non-linear resistive device, the value of resistance would change with operating conditions. The volt-ampere characteristic curve for such a device is shown in Figure 14-2. This non-linear curve displays unequal changes in the dependent variable (power) for equal changes in the independent variable (current). As is customary, the independent variable is plotted on the horizontal axis and, in this case, is expressed in milliamperes. The change in ΔI_1 is equal to the change in ΔI_2, but a comparison of ΔP_1 and ΔP_2 clearly shows they are not equal.

NOTE: Both linear and non-linear volt-ampere (E-I) characteristic curves are widely used in electronics. From either of these curves it is possible to determine both the static and dynamic values of resistance.

STATIC RESISTANCE

Static resistance is DC resistance. It is determined by selecting a voltage point, see Figure 14-3, and determining the associated current point. The voltage is then divided by the current to obtain resistance.

EXAMPLE 1:

What is the static resistance at (1) the 3-volt point, and (2) the 6-volt point for the curve of Figure 14-3?

SOLUTION: First, draw in the dashed lines. Observe how a perpendicular line is drawn upward from the voltage axis until it intersects with the curve. Another line is then drawn perpendicular to the current axis which must also intersect the curve at the same point. The 3-volt line is seen to intersect the curve at the point where the current is 1.5 mA. Thus, the static resistance is determined by Ohm's law.

$$\frac{3\,V}{1.5\,mA} = 2,000 \text{ ohms}$$

Repeating this procedure for the 6-volt point shows the line to intersect the curve at the point where the current is 3 mA. Again,

$$\frac{6\,V}{3\,mA} = 2,000 \text{ ohms}$$

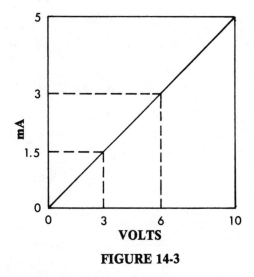

FIGURE 14-3

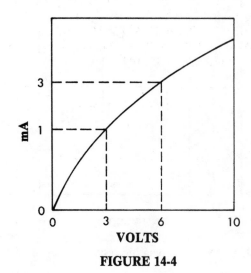

FIGURE 14-4

EXAMPLE 2:

Determine the static resistance for (1) the 3-volt point of Figure 14-4 and (2) the 6-volt point.

SOLUTION: For the 3-volt point, the resistance is

$$\frac{3\,V}{1\,mA} = 3,000 \text{ ohms}$$

For the 6-volt point, the resistance is:

$$\frac{6\,V}{3\,mA} = 2,000 \text{ ohms}$$

Further investigation of the curve, Figure 14-4, would show that as the voltage is increased the static value of resistance would decrease.

DYNAMIC RESISTANCE

Dynamic resistance is determined by dividing the incremental change in voltage (ΔE) by the incremental change in current (ΔI). It will be observed that the dynamic resistance can be less than, equal to, or greater than the static resistance depending on the slope of the curve.

EXAMPLE 3:
Determine the dynamic resistance for (1) the change ΔE_1, area A, and (2) the change ΔE_2, area B, Figure 14-1.

   SOLUTION: The dynamic resistance for area A is

$$\frac{\Delta E_1}{\Delta I_1} = \frac{2-1}{4-2} = \frac{1 \text{ V}}{2 \text{ mA}} = 500 \text{ ohms}$$

The dynamic resistance for area B is

$$\frac{\Delta E_2}{\Delta I_2} = \frac{7-6}{14-12} = \frac{1 \text{ V}}{2 \text{ mA}} = 500 \text{ ohms}$$

EXAMPLE 4:
   Determine the dynamic resistance for (1) the change ΔE_1, area A, and (2) the change ΔE_2, area B, of Figure 14-5.

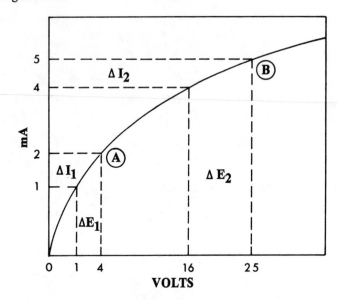

FIGURE 14-5

SOLUTION: The dynamic resistance for area A is

$$\frac{\Delta E_1}{\Delta I_1} = \frac{4-1}{2-1} = \frac{3 \text{ V}}{1 \text{ mA}} = 3,000 \text{ ohms}$$

The dynamic resistance for area B is

$$\frac{\Delta E_2}{\Delta I_2} = \frac{25-16}{5-4} = \frac{9 \text{ V}}{1 \text{ mA}} = 9,000 \text{ ohms}$$

SUGGESTED OBSERVATIONS

1. Include all tables and graphs in the experimental write-up.
2. Handle components with care, as leads are easily broken.
3. Don't exceed the recommended current and voltage ratings given in the experiment.
4. Record all measurements and calculations in the data table.
5. Review graphing techniques: Appendix C-1.

LIST OF MATERIALS

1. VPS
2. VOM
3. MFM
4. No. 1819 Lamp with holder
5. CL704L Photoconductive Cell
6. 1N4001 Diode
7. 1 kΩ, 2 watt Resistor
8. 41D2 Thermistor
9. Three 1.5 V cells
10. 1 kΩ Potentiometer
11. Graph Paper (10 × 10 to the inch)

Alternate Materials

_____

_____

_____

_____

_____

EXPERIMENTAL PROCEDURE

The following four electronic devices will be examined in this experiment for linearity and bilateral characteristics.

FIGURE 14-6

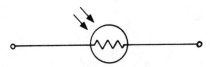

FIGURE 14-7

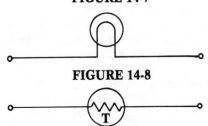

FIGURE 14-8

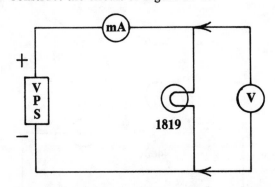

FIGURE 14-9

The DIODE symbol is shown in Figure 14-6. The A indicates the anode and it is the electron catching element. The K is the cathode or electron emitting element. Generally, the symbol is stamped on metal or plastic bodies. For some diodes, the cathode is the end with the color bands.

The symbol for the PHOTOCONDUCTIVE DEVICE (Photoresistor) is shown in Figure 14-7. The device changes its resistance when light strikes the photoresistive surface.

The INCANDESCENT LAMP symbol is shown in Figure 14-8.

The symbol for the THERMISTOR is shown in Figure 14-9. The resistance of the thermistor changes with changes in temperature.

SECTION A: TUNGSTEN LAMP CHARACTERISTICS

1. Construct the circuit of Figure 14-10.

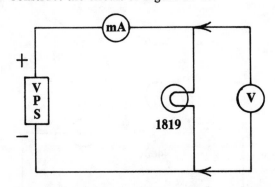

FIGURE 14-10

2. Adjust the VPS for a reading of 5 mA and measure the voltage dropped across the lamp. Repeat this procedure in 5 mA increments from 5 mA to 40 mA. Construct a table for values of current versus voltage.
3. Using the values determined in Step 2, plot a curve of current versus voltage.
4. Is the curve of Step 3 linear?

SECTION B: PHOTOCONDUCTIVE CELL CHARACTERISTICS

1. Construct the circuit of Figure 14-11. Place the lamp and cell in a light tight tube.

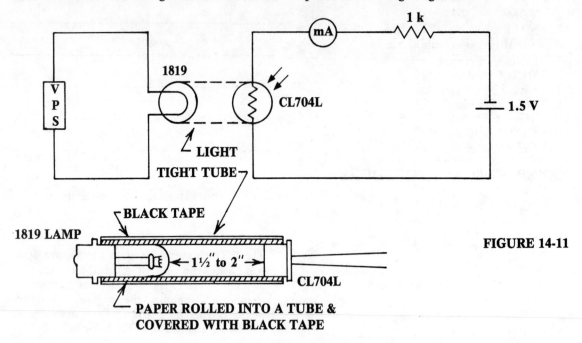

FIGURE 14-11

2. The intensity of the lamp can be controlled by varying the VPS. This in turn will cause the resistance of the photoconductive cell to vary. Adjust the VPS until 1 mA is indicated by the milliammeter. Place an electronic MFM across the photoconductive cell and record the voltage.

3. Reduce the light intensity by decreasing the VPS until the milliammeter reads 0.9 mA. Record the voltage across the cell. Repeat this procedure in 0.1 mA increments until zero current is indicated by the milliammeter. Construct a table of values for the voltage across versus the current through the photoconductive cell.

4. Using the values obtained in Step 3, plot a curve of current versus voltage.

5. Compute the resistance of the photoconductive cell when 1 mA is flowing in the circuit.

6. Compute the resistance of the photoconductive cell when 0.1 mA is flowing in the circuit.

7. Is the graph of Step 4 a linear curve?

8. Using the same circuit (Figure 14-11), construct a table of values for the voltage across the lamp and the current through the photoconductive cell. Allow the current through the cell to vary from 0 to 1 mA in 0.1 mA increments.

9. Plot a curve using the values of voltage and current determined in Step 8. Plot the current through the cell versus the voltage drop across the light source.

10. Is the graph of Step 9 a linear curve?

SECTION C: DIODE CHARACTERISTICS

1. Construct the circuit of Figure 14-12(a).

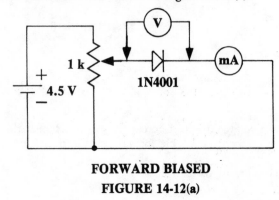

FORWARD BIASED

FIGURE 14-12(a)

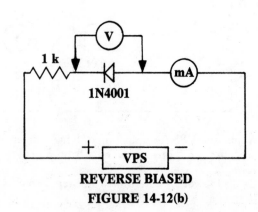

REVERSE BIASED

FIGURE 14-12(b)

2. Starting with a voltmeter reading of zero volts and progressing in 0.1 volt increments, record the milliammeter reading. Do not exceed 40 mA. Construct a table of values for current versus voltage.
3. Construct the circuit of Figure 14-12(b).
4. Starting with a voltmeter reading of zero volts and progressing in 5-volt increments, record the milliammeter reading. Do not exceed 40 volts. Construct a table of values for current versus voltage.
5. On a piece of linear graph paper, construct and label the axes for a graph of the diode curve. Use Figure 14-13 as an example.
6. Using the table of values of Step 2, plot the diode's forward bias characteristics.
7. Using the table of values of Step 4, plot the diode's reverse bias characteristics.
8. From the curve, determine the static resistance for 0.4 volts of forward voltage.
9. From the curve, determine the static resistance for 40 volts of reverse voltage.
10. Using the delta process, approximate the slope at three different locations on the curve of Steps 6 and 7 by fitting straight lines to it. See Figure 14-14 for the technique of expanding the straight line formed by the delta.
11. Using the modified curve of Step 10, determine the dynamic resistance for a 0.1 forward voltage change between 0.3 and 0.4 volts.

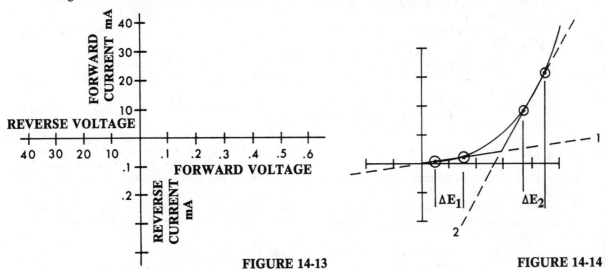

FIGURE 14-13   FIGURE 14-14

SECTION D: THERMISTOR CHARACTERISTICS
1. Construct the circuit of Figure 14-15.

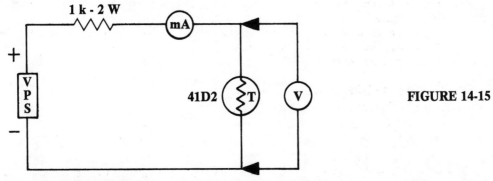

FIGURE 14-15

2. While making the following measurements, protect the thermistor from air movement within the room to prevent erroneous readings. In addition, allow 30 to 40 seconds for the thermistor to stabilize before taking the readings.
3. Starting with 1 mA and moving in 2 mA increments up to a maximum of 35 mA, construct a table of values by measuring the voltage across and the current through the thermistor of Figure 14-15.
4. Plot a curve of voltage versus current for the values determined in Step 3.
5. Using the curve, determine the resistance of the thermistor when 1 mA is flowing in the circuit.
6. Repeat Step 5 for 35 mA.
7. From the curve, determine the dynamic resistance in the 15 to 20 mA region.

103

SECTION E: COMPONENT CHARACTERISTICS

Use the term linear bilateral, linear unilateral, non-linear bilateral, or non-linear unilateral to describe the diode, lamp, photoconductive cell, and thermistor.

1. Choose a component. Using the ohmmeter function of the MFM determine whether the device is bilateral or unilateral. To do this, first measure the resistance of the component, then reverse the leads and again measure the resistance. If the resistance reading is the same in both directions, then the device is bilateral; otherwise, it is unilateral.
2. Determine if the device is linear or non-linear by reviewing the previously plotted E-I curve of the component.
3. Describe the device as linear bilateral, linear unilateral, non-linear bilateral, or non-linear unilateral.
4. Repeat the procedure for each component.

DATA INTERPRETATION AND CONCLUSIONS

Write a general summary of the ideas presented in this experiment. This discussion should include:
1. A comparison of the E-I curve characteristic of the components investigated in this experiment.
2. A brief description of each component's electrical characteristic.
3. Your own conclusions.

APPLICATIONS

1. Diodes: (a) power supplies, (b) computers, (c) radios, (d) televisions, and (e) telephone circuits.
2. Photoconductive Cells: (a) optical link, (b) automatic door openers, (c) burglar alarms, (d) computers, and (e) card readers.
3. Thermisters: (a) resistive thermometers, (b) liquid level indicators, (c) anemometers, (d) gas analysers, and (e) temperature compensators.

PROBLEMS

1. Use a parts catalog to determine the cost of the lamp, diode, thermistor, and photoconductive cell used in this experiment.
2. If the photoconductive cell has a maximum power dissipation of 100 mW, what is the maximum current that can safely be handled when the resistance of the photoconductive cell is 1000 ohms?

15 | MILLMAN'S THEOREM

OBJECTIVES

Millman's Theorem provides a method of reducing a network of two or more sources to a single source for purposes of circuit analysis. Ideally, a complex circuit can be reduced to a Thevenin equivalent using this theorem. This experiment points out the advantage of Millman's theorem when certain types of circuits are to be analyzed.

MILLMAN'S THEOREM

The use of Millman's theorem lends itself to applications where it is desirable to determine the voltage across two points in a network. It can be shown to be an evolution of Norton's concepts. To illustrate, consider the network of Figure 15-1. As has been previously shown, this network can be reduced to several

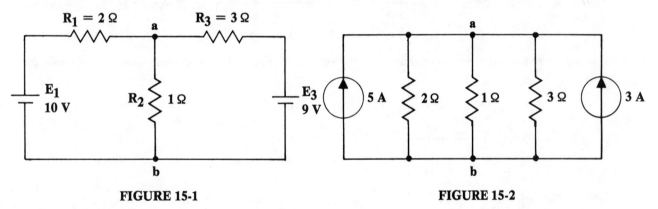

FIGURE 15-1   FIGURE 15-2

Norton equivalents as in Figure 15-2. The short circuit current of the 10 volt, 2 ohm source is expressed mathematically as

$$\frac{E_1}{R_1} = \frac{10\ V}{2\ \Omega} = 5 \text{ amperes.}$$

The short circuit current of the 0 volt, 1 ohm source is

$$\frac{E_2}{R_2} = \frac{0\ V}{1\ \Omega} = 0 \text{ amperes}$$

And the short circuit current of the 9 volt, 3 ohm source is

$$\frac{E_3}{R_3|} = \frac{9\ V}{3\ \Omega} = 3 \text{ amperes}$$

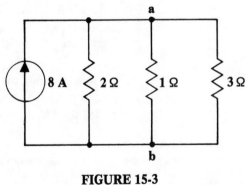

FIGURE 15-3

These currents are then added algebraically and indicated as a single current source as shown in Figure 15-3. Therefore, the total current in Figure 15-2 is:

$$\frac{E_1}{R_1} + \frac{E_2}{R_2} + \frac{E_3}{R_3} = 5\ A + 0\ A + 3\ A = 8\ A$$

Combining the conductances of the 2, 1, and 3 ohm resistances, results in:

$$G_T = G_1 + G_2 + G_3 = \frac{1}{2} + \frac{1}{1} + \frac{1}{3} = 1.833 \text{ siemens}$$

as shown in Figure 15-4. Using Equations 5-4(a) and 5-5(a), the Norton's equivalent circuit of Figure 15-4 can be converted to the Thevenin's equivalent circuit of Figure 15-5. Thus E_{ab} is 4.36 volts.

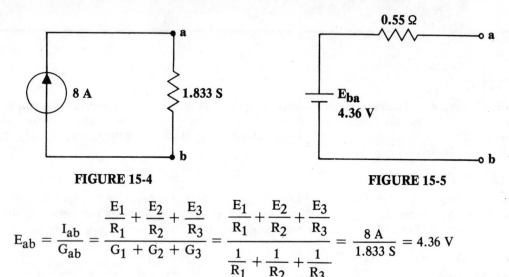

FIGURE 15-4   **FIGURE 15-5**

$$E_{ab} = \frac{I_{ab}}{G_{ab}} = \frac{\dfrac{E_1}{R_1} + \dfrac{E_2}{R_2} + \dfrac{E_3}{R_3}}{G_1 + G_2 + G_3} = \frac{\dfrac{E_1}{R_1} + \dfrac{E_2}{R_2} + \dfrac{E_3}{R_3}}{\dfrac{1}{R_1} + \dfrac{1}{R_2} + \dfrac{1}{R_3}} = \frac{8 \text{ A}}{1.833 \text{ S}} = 4.36 \text{ V}$$

From the foregoing it can be seen that if the voltage E_{ab} in Figure 15-1 is desired, then the equation for the solution is:

$$E_{ab} = \frac{\dfrac{E_1}{R_1} + \dfrac{E_2}{R_2} + \dfrac{E_3}{R_3}}{\dfrac{1}{R_1} + \dfrac{1}{R_2} + \dfrac{1}{R_3}}$$

and Millman's Equation is:

$$E = \frac{\dfrac{E_1}{R_1} + \dfrac{E_2}{R_2} + \dfrac{E_3}{R_3} + ... + \dfrac{E_n}{R_n}}{\dfrac{1}{R_1} + \dfrac{1}{R_2} + \dfrac{1}{R_3} + ... + \dfrac{1}{R_n}}$$

EQUATION 15-1

where: E = voltage across the two points. Be sure to include the reference subscript
E_1, E_2, E_3, etc. = voltage in each branch with polarity determined from reference point
R_1, R_2, R_3, etc. = resistance in each branch

EXAMPLE 1:

Determine the voltage E_{ag} for the circuit of Figure 15-6.

SOLUTION: Using "g" as the reference point, the voltage of E_{ag} is

$$E_{ag} = \frac{\dfrac{100 \text{ V}}{700 \text{ k}\Omega} + \dfrac{0}{1 \text{ k}\Omega} + \dfrac{-50 \text{ V}}{560 \text{ k}\Omega}}{\dfrac{1}{700 \text{ k}\Omega} + \dfrac{1}{1 \text{ k}\Omega} + \dfrac{1}{560 \text{ k}\Omega}} = \frac{\dfrac{1}{7} - \dfrac{0.5}{5.6}}{\dfrac{1}{700} + \dfrac{1}{1} + \dfrac{1}{560}}$$

$$E_{ag} = \frac{0.143 - 0.0893}{0.00143 + 1 + 0.00179} = \frac{0.0537}{1.00322} = 53.5 \text{ mV}$$

FIGURE 15-6

EXAMPLE 2:

Determine the voltage E_{ag} for the circuit of Figure 15-7

SOLUTION: Using "g" as the reference point, the voltage of E_{ag} is

$$E_{ag} = \frac{\dfrac{35\ V}{680\ k\Omega} - \dfrac{50\ V}{560\ k\Omega}}{\dfrac{1}{680\ k\Omega} + \dfrac{1}{10\ M\Omega} + \dfrac{1}{560\ k\Omega}} = \frac{0.0515 - 0.0893}{0.00147 + 0.0000001 + 0.00179}$$

$$E_{ag} = \frac{-0.0378}{0.00326} = -11.6\ V$$

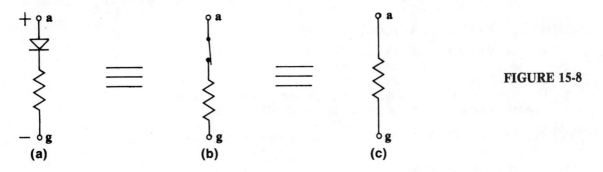

FIGURE 15-7

Suppose a diode has an internal resistance of 1000 ohms when the diode is conducting. Such a condition is shown in Figure 15-8.

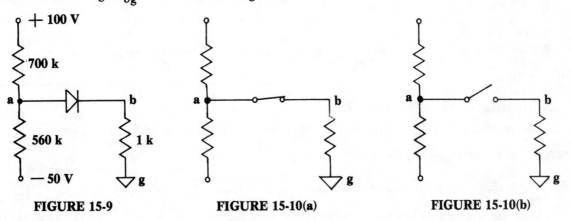

FIGURE 15-8

(a)   (b)   (c)

EXAMPLE 3:

Determine the voltage E_{bg} for the circuit of Figure 15-9.

FIGURE 15-9   FIGURE 15-10(a)   FIGURE 15-10(b)

SOLUTION: If the voltage E_{ag} is positive, the diode will conduct and the circuit becomes identical to the circuit of Figure 15-10(a). On the other hand, if the voltage E_{ag} is negative, the circuit will appear as shown in Figure 15-10(b). To determine the polarity of E_{ag}, temporarily remove the diode from the circuit, see Figure 15-11 and use Millman's equation to compute E_{ag}. It can be seen that the voltage is positive. Therefore, the circuit is the same as Figure 15-6 where E_{bg} equals 53.5 mV.

$$E_{ag} = \frac{\dfrac{100\ V}{700\ k\Omega} - \dfrac{50\ V}{560\ k\Omega}}{\dfrac{1}{700\ k\Omega} + \dfrac{1}{560\ k\Omega}} = +16.7\ V$$

The results of having a negative voltage, E_{ag}, can be shown with Figure 15-12. Observe that it is

107

identical to the circuit of Figure 15-7. Removal of the diode and computing using Millman's theorem shows that E_{ag} is a negative 11.78 V. Therefore, the diode looks like an open switch and thus a high resistance such as the 10 meg ohm resistor of Figure 15-12.

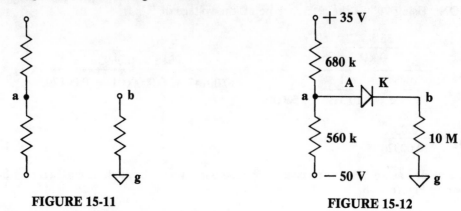

FIGURE 15-11   **FIGURE 15-12**

SUGGESTED OBSERVATIONS

1. Use the % error equation (Equation 4-7) to compare values.
2. Complete calculation prior to performing the experiment.
3. Handle components with care.
4. Do not exceed the recommended voltages as components may be damaged.
5. Record all measurements and calculations in the data tables.

LIST OF MATERIALS

1. MFM
2. Two 22.5 - 45 volt "B" Batteries
3. Four 1.5 V Cells
4. Thermistor 41D2
5. Photoconductive Cell CL704L
6. Lamp No. 1819
7. Diode IN4001
8. Resistors—all ½ watt except where noted

| 220 Ω | 470 Ω | 4.7 kΩ (1 W) |
| 390 Ω | 1 kΩ | 10 kΩ |

Alternate Materials

2 Regulated VPS

_____

_____

_____

_____

EXPERIMENTAL PROCEDURE

This experiment is design to provide you with a working knowledge of Millman's theorem.

SECTION A: MILLMAN'S THEOREM

1. Using Millman's theorem, determine E_{ab}.
2. Measure E_{ab}.
3. Compare E_{ab} of Step 1 to that of Step 2.

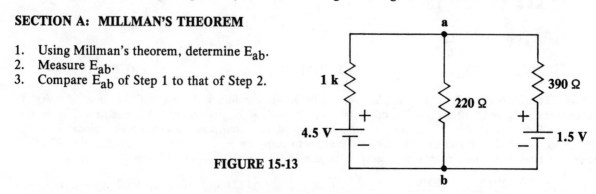

FIGURE 15-13

SECTION B: APPLICATION OF MILLMAN'S THEOREM

1. With the MFM, measure the cold resistance of the thermistor (approximately 10 kΩ).
2. Since it is unknown whether the diode will conduct, draw two equivalent circuits of Figure 15-14 with one showing the diode with zero resistance (diode conducting) and the other showing the diode with infinite resistance (diode not conducting).

3. Using Millman's theorem, determine the voltage drop E_{ag} for each of the circuits of Step 2.
4. Will the diode conduct? (Anode polarity in relation to the cathode as specified in Experiment 14, Figure 14-6.)
5. Repeat Steps 2 through 4 for a thermistor resistance of 1 kΩ.
6. Will the diode conduct?

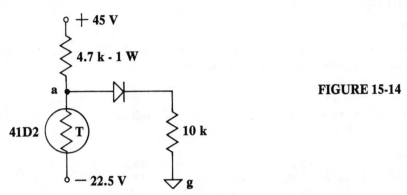

FIGURE 15-14

7. Construct the circuit of Figure 15-14. Use 45 V batteries or regulated power supplies for the source voltages. Replace the thermistor with a value of resistance similar to that measured in Step 1 (about 10 kΩ). Measure E_{ag}.
8. Compare the value of E_{ag} of Step 7 to that voltage calculated in Step 3 which indicated the state of the diode as described in Step 4.
9. Repeat Step 7; this time replace the thermistor with a 1 kΩ resistor. Measure E_{ag}.
10. Compare the value of E_{ag} of Step 9 with that voltage calculated in Step 5 which indicated the state of the diode as described in Step 6.
11. Place the thermistor in the circuit of Figure 15-14. Monitor the voltage from **a** to **g** and observe the voltage change as the thermistor changes resistance. Start with the thermistor in a cold state. Once E_{ag} has changed polarity, blow across the thermistor. Record your observation.

SECTION C: APPLICATION OF MILLMAN'S THEOREM

1. Assuming the CL704L has a "dark" resistance of 400 kΩ and a "light" resistance of 0.6 kΩ, draw two equivalent circuits (one for dark conditions and the other for light conditions) of Figure 15-15.
2. Using Millman's theorem, determine E_{ag} for each condition.

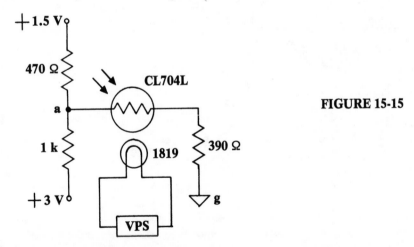

FIGURE 15-15

3. Construct the circuit of Figure 15-15. Place the cell in a light tight tube with a No. 1819 lamp as was done in Experiment 14, Section B.
4. Measure E_{ag} with the light turned off.
5. Compare E_{ag} of Step 4 to that calculated in Step 2.
6. Place 18 to 23 V across the lamp and measure E_{ag}.
7. Compare E_{ag} of Step 6 to that calculated in Step 2.

DATA INTERPRETATION AND CONCLUSIONS

Write a general summary of the ideas presented in this experiment. This discussion should include:
1. The usefulness of Millman's theorem.
2. The validity of the answers arrived at when compared to measured circuit conditions.
3. Your own conclusions.

APPLICATIONS

Millman's theorem finds application (1) when Thevenizing a network, (2) in vacuum tube and transistor circuit analysis, (3) when the voltage between two points is required.

PROBLEM

1. The lamp in the following circuit lights when a voltage of 28 volts and a current of 300 mA excites it. Determine if the lamp will light.

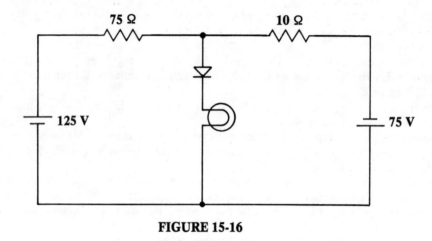

FIGURE 15-16

16 | LOOPS AND NODES

OBJECTIVES

Loop and nodal equations, based on Kirchhoff's voltage and current laws, were written and applied in earlier experiments to solve comparatively simple circuit problems. In this experiment, a systematic method is developed for using **loop equations** to determine the current at any point in a network. A systematic method is also developed for using **nodal equations** to determine the voltage across any node in a network. Using these methods, you will solve for voltage and current in several networks and then you will check out these calcuations.

CIRCUIT ANALYSIS USING LOOP EQUATIONS

The development of the loop equation is based upon the premise that the algebraic sum of the voltages around a closed loop is equal to zero—a concept first presented in Experiment 4. The unknown quantity in the loop equation is the **loop current**. Several simple rules must be observed when writing loop equations. Using Figure 16-1 as an illustration, these are the rules:

1. **Draw a network graph of the original circuit.** This consists of an outline of the drawing without the circuit elements shown. See Figure 16-2.

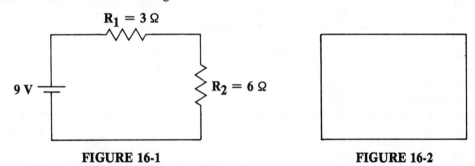

| FIGURE 16-1 | FIGURE 16-2 |

2. **Determine the number of closed loops in the network graph. The number of closed loops indicates the number of loop equations required to solve the network.** Figure 16-2 shows only one loop, hence only one loop equation is required.
3. **Draw in a current loop in the network graph and then draw in the same current loop within the original circuit.** The direction of the current loop can be drawn either clockwise or counterclockwise. See Figure 16-3.

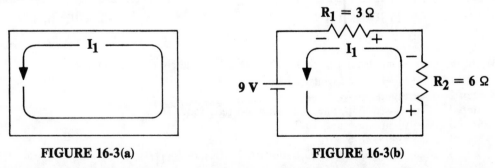

| FIGURE 16-3(a) | FIGURE 16-3(b) |

4. **Assume that the direction of the current loop is correct and mark each resistor with the polarity that the current loop would produce as it passes through each resistor.** Once you understand the mechanics of writing loop equations, this step may be omitted.

5. The loop equation for one loop is Equation 16-1.

$$K_1 = R_{11}I_1$$ **EQUATION 16-1**

   where: K_1 = the algebraic sum of all the voltage sources that the assumed loop
              current (I_1) passes through. The subscript 1 identifies the loop that
              is being worked with.

   R_{11} = the **mesh** or **loop** resistance. The first subscript identifies the loop
              that is being worked with while the second indicates the current of
              the loop number. As an example, the following question should be
              asked: "What resistors in loop 1 (first subscript) are effected by the
              current of loop 1 (second subscript)?" Thus, the mesh resistance of
              R_{11} is the sum of all the resistors that the current (I_1) is assumed to
              be flowing through.

   I_1 = the assumed current for loop number 1.

6. **Starting at the head and moving toward the tail of the current loop, the polarity and magnitude of K_1 is determined as follows: if the first terminal of the source encountered is positive, then the magnitude of the source is positive.** Thus, the first, and only, source encountered in Figure 16-3 has a magnitude of 9 volts with the first terminal being approached from the negative side. Therefore, $K_1 = -9$.

7. Following the current loop No. 1 from head to tail, add up each resistance that the loop goes through. This will be the value of R_{11}. Thus for Figure 16-3, the current loop goes through a 3 ohm and a 6 ohm resistor and $R_{11} = 3 + 6 = 9$.

8. Substituting the values determined in Steps 6 and 7 into Equation 16-1 (Step 5) results in

$$K_1 = R_{11}I_1$$

$$-9 = 9I_1$$

$$I_1 = -1 \text{ ampere}$$

The **NEGATIVE** sign indicates that the **assumed** current direction was wrong.

9. **Redraw** the circuit of Figure 16-3 and then draw in the current loop opposite to the direction originally assumed. See Figure 16-4. The actual current can then be shown as illustrated in Figure 16-5.

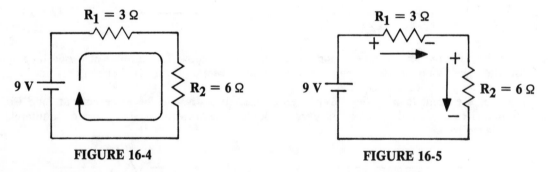

FIGURE 16-4   **FIGURE 16-5**

EXAMPLE 1:

Write the loop equation for Figure 16-6. Note this is the circuit of Figure 4-2.

SOLUTION: Follow the steps just outlined.

Step 1. See Figure 16-7.
Step 2. One closed loop, one loop equation required.
Step 3. See Figure 16-6 and 16-7.
Step 4. See Figure 16-6.
Step 5. Equation required is Equation 16-1. $(K_1 = R_{11}I_1)$
Step 6. $K_1 = 3 - 8 - 10 = -15$.
Step 7. $R_{11} = 6 + 7 + 2 = +15$.
Step 8. $-15 = 15I_1$. Thus $I_1 = -1$ ampere.

112

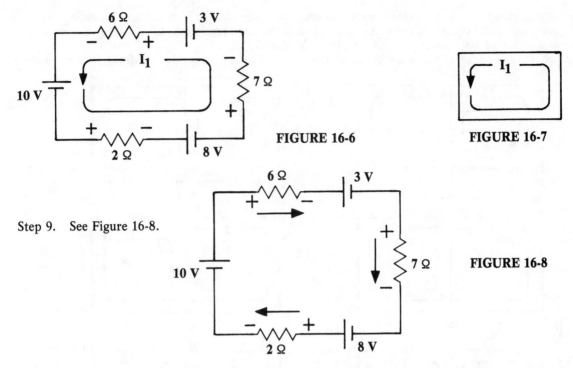

FIGURE 16-6

FIGURE 16-7

Step 9.   See Figure 16-8.

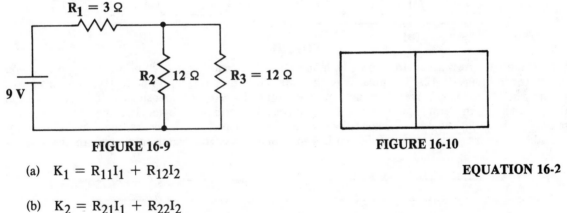

FIGURE 16-8

Figure 16-1 is reconfigured as shown in Figure 16-9. The network graph of Figure 16-9 is shown in Figure 16-10. Note that there are two closed loops. Thus, two loop equations are required for the solution of the loop currents I_1 and I_2. These equations are written as follows:

FIGURE 16-9

FIGURE 16-10

(a) $K_1 = R_{11}I_1 + R_{12}I_2$

EQUATION 16-2

(b) $K_2 = R_{21}I_1 + R_{22}I_2$

where: K_1, R_{11}, and I_1 are the same as in Equation 16-1.

K_2 =   the algebraic sum of all the voltage sources that the assumed loop current (I_2) passes through. The subscript 2 identifies the loop that is being worked with.

R_{12} =   the mutual resistance between loops 1 and 2. This is the resistance **shared** by other loop currents. Again the question must be asked: "What resistors in loop 1 (first subscript) are effected by, or share, the current of loop 2 (second subscript)?"

R_{21} =   the mutual resistance between loops 2 and 1. The question is asked: "What resistors in loop 2 are effected by, or share, the current of loop 1?"

R_{22} =   identical to the statements for R_{11} except that the subscripts have been changed. The question is: "What resistors in loop 2 are effected by the current of loop 2?"

113

When two or more unknown currents interact, their solution is obtained with simultaneous equations. Because loop and nodal equations are easily adapted to the determinant methods of solving simultaneous equations, a review of determinants is provided in Appendix C-2.

Figure 16-11 shows a few of the many ways that the assumed current loops can be drawn. Observe that

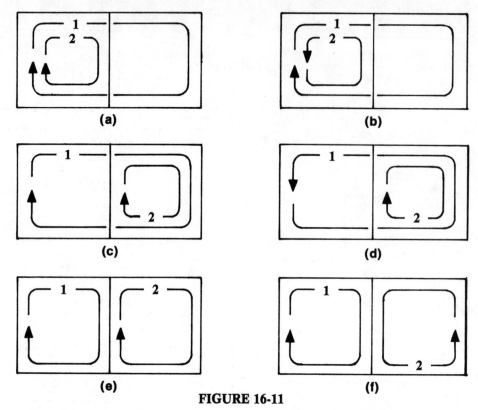

FIGURE 16-11

a few "mutual" resistances are shown with some loop currents aiding while others oppose, a concept illustrated in Figure 16-12. The polarity of the "mutual term", e.g. $R_{12}I_2$ or $R_{21}I_1$, is determined after noting whether the loop currents are aiding, as in Figure 16-12(a), or opposing, as in Figure 16-12(b). If the currents aid, the term is positive and if they oppose, the term is negative. In some circuits it is possible to have both aiding and opposing conditions for the same loops. If this occurs, then the individual terms are added algebraically. An example that has this condition is given later on in this experiment.

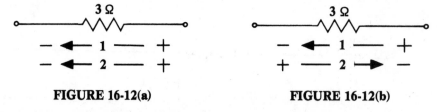

FIGURE 16-12(a)   **FIGURE 16-12(b)**

A considerable amount of the labor required to solve circuit currents can be reduced if the loops are drawn with a specific plan in mind. As an example, consider the circuit of Figure 16-13.

EXAMPLE 2:

Determine the magnitude and direction of the current flowing in R_2 which is the 2 ohm resistor of Figure 16-13.

    SOLUTION: Follow the steps previously outlined.
    Step 1.   See Figure 16-14.
    Step 2.   Two closed loops, two equations required.
    Step 3.   Since the current to be determined is in the center leg, only one loop (No. 2) will be drawn
              through it so that only one unknown (I_2) will have to be solved for.

Step 4. No longer required because of the rule on aiding and opposing currents in the mutual resistances. All **mesh** resistances are positive.

Step 5. Equations required are Equations 16-2(a) and (b).

(a) $K_1 = R_{11}I_1 + R_{12}I_2$

(b) $K_2 = R_{21}I_1 + R_{22}I_2$

Step 6 and 7.

$$K_1 = (-5) + (-10) = -15$$

$$R_{11} = (+2) + (+1) = +3$$

$$R_{12} = +2 \quad \text{(This is } R_1.)$$

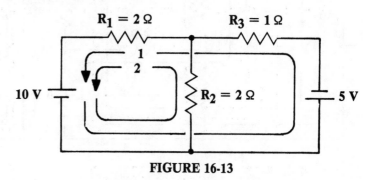

FIGURE 16-13

Step 8(c). Thus,

$$-15 = 3I_1 + 2I_2$$

Step 6 and 7(b).

$$K_2 = -10$$

$$R_{21} = +2 \quad \text{(This is } R_1.)$$

$$R_{22} = (+2) + (+2) = +4$$

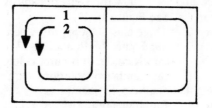

FIGURE 16-14

Step 8(b). Thus,

$$-10 = 2I_1 + 4I_2$$

Step 8(c).

$$-15 = 3I_1 + 2I_2$$

$$-10 = 2I_1 + 4I_2$$

Solving by determinants, the denominator is reduced to

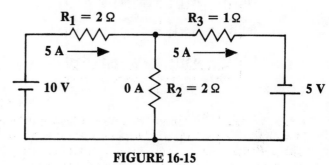

FIGURE 16-15

$$\Delta = \begin{vmatrix} 3 & 2 \\ 2 & 4 \end{vmatrix} = 12 - 4 = 8$$

$$I_2 = \frac{\begin{vmatrix} 3 & -15 \\ 2 & -10 \end{vmatrix}}{\Delta} = \frac{(-30) + 30}{8} = 0 \text{ amperes}$$

Step 9. See Figure 16-15.

Although the example did not specifically request that I_1 be determined, it will be derived to show how the currents going through R_1 and R_3 are handled. No current flows through R_2, but a current of 5 amperes is flowing in R_1 and R_3—and in a direction opposite to that initially assumed.

$$I_1 = \frac{\begin{vmatrix} -15 & 2 \\ -10 & 4 \end{vmatrix}}{\Delta} = \frac{(-60) + 20}{8} = -5 \text{ amperes}$$

EXAMPLE 3:

Determine both magnitude and direction of current flow in the 4-ohm resistor shown in the circuit of Figure 16-16.

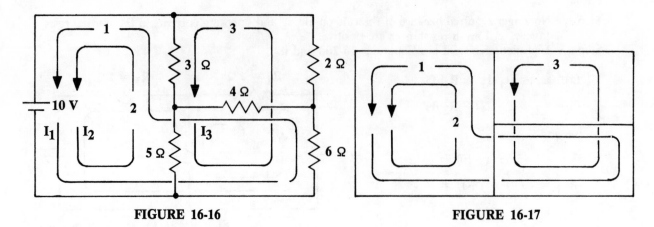

FIGURE 16-16

FIGURE 16-17

SOLUTION: Follow the steps previously outlined.

Step 1. See Figure 16-17.

Step 2. Three closed loops, three equations are required.

Step 3. See Figure 16-16 and 16-17. Again, there are many ways that the loops may be drawn. In this example, only current loop number 1 is drawn through the 4-ohm resistor. This reduces the number of solutions required.

Step 4. See Step 4, Example 2.

Step 5. Equations required are Equation 16-3(a), (b), and (c).

(a) $K_1 = R_{11}I_1 + R_{12}I_2 + R_{13}I_3$ **EQUATION 16-3**

(b) $K_2 = R_{21}I_1 + R_{22}I_2 + R_{23}I_3$

(c) $K_3 = R_{31}I_1 + R_{32}I_2 + R_{33}I_3$

Step 6 and (7a).

$K_1 = -10$

$R_{11} = 3 + 4 + 6 = 13$

$R_{12} = +3$

$R_{13} = (-3) + 6 = 3$ (See Figure 16-18)

Step 8(a). Thus, $-10 = 13I_1 + 3\,I_2 + 3I_3$

Step 6 and 7(b).

$K_2 = -10$

$R_{21} = +3$

$R_{22} = 3 + 5 = +8$

$R_{23} = (-3) + (-5) = -8$

Step 8(b). Thus, $-10 = 3I_1 + 8I_2 - 8I_3$

Step 6 and 7(c).

$K_3 = 0$

$R_{31} = (-3) + 6 = 3$

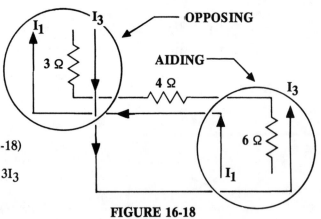

FIGURE 16-18

116

$$R_{32} = (-3) + (-5) = -8$$

$$R_{33} = 2 + 6 + 5 + 3 = 16$$

Step 8(c). Thus, $0 = 3I_1 - 8I_2 + 16I_3$

Step 8(d).

$$-10 = 13I_1 + 3I_2 + 3I_3$$

$$-10 = 3I_1 + 8I_2 - 8I_3$$

$$0 = 3I_1 - 8I_2 + 16I_3$$

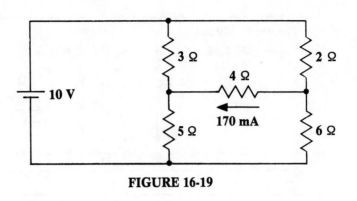

FIGURE 16-19

Solving by determinants, the denominator is reduced to

$$\Delta = \begin{vmatrix} 13 & 3 & 3 \\ 3 & 8 & -8 \\ 3 & -8 & 16 \end{vmatrix} \begin{matrix} 13 & 3 \\ 3 & 8 \\ 3 & -8 \end{matrix}$$

$$\Delta = (13 \times 8 \times 16) + (3 \times -8 \times 3) + (3 \times 3 \times -8) - (3 \times 8 \times 3) - (-8 \times -8 \times 13) - (16 \times 3 \times 3)$$

$$\Delta = 472$$

$$I_1 = \dfrac{\begin{vmatrix} -10 & 3 & 3 \\ -10 & 8 & -8 \\ 0 & -8 & 16 \end{vmatrix} \begin{matrix} -10 & 3 \\ -10 & 8 \\ 0 & -8 \end{matrix}}{\Delta}$$

$$I_1 = \frac{(-10 \times 8 \times 16) + (3 \times -8 \times 0) + (3 \times -10 \times -8) - (0 \times 8 \times 3) - (-8 \times -8 \times -10) - (16 \times -10 \times 3)}{472}$$

$$I_1 = \frac{80}{472} = 0.17 \text{ A}$$

Step 9. See Figure 16-19.

EXAMPLE 4:

Determine currents I_2 and I_3 for the circuit of Figure 16-16 and include both magnitude and direction for all currents.

SOLUTION:

$$I_2 = \dfrac{\begin{vmatrix} 13 & -10 & 3 \\ 3 & -10 & -8 \\ 3 & 0 & 16 \end{vmatrix} \begin{matrix} 13 & -10 \\ 3 & -10 \\ 3 & 0 \end{matrix}}{\Delta} = \frac{-1270}{472} = -2.69 \text{ A}$$

$$I_3 = \dfrac{\begin{vmatrix} 13 & 3 & -10 \\ 3 & 8 & -10 \\ 3 & -8 & 0 \end{vmatrix} \begin{matrix} 13 & 3 \\ 3 & 8 \\ 3 & -8 \end{matrix}}{\Delta} = \frac{-650}{472} = -1.38 \text{ A}$$

All currents have been shown with magnitude and direction in Figure 16-20. Figure 16-21 shows the final current pattern after each current has been summed up algebraically.

117

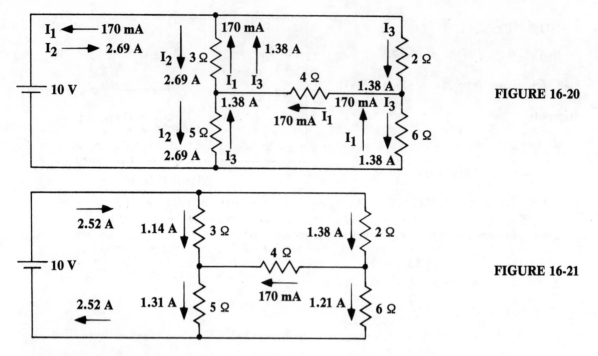

FIGURE 16-20

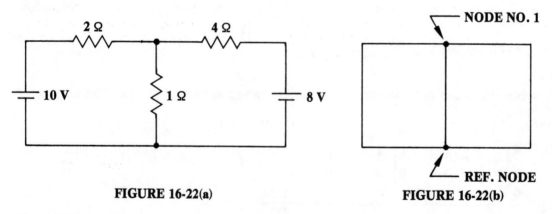

FIGURE 16-21

As a final check, use Kirchhoff's current law at each node to see if the currents leaving equal the currents entering.

NODAL ANALYSIS USING NODAL EQUATIONS

Nodal equations were developed around the concept that the algebraic sum of all currents entering a node must equal zero. The unknown quantity in the **nodal equation** is the **node voltage**.

Recalling that a **major node** is the connecting point of three or more elements, it can be shown that network graphs with two or more loops will also have two or more nodes. One node is selected as the reference node. The remaining nodes indicate the number of nodal equations required to solve the network and are labeled Node No. 1, Node No. 2, etc. Thus, Figure 16-22 has two major nodes. One will be selected as the reference node and one will be identified as Node No. 1 to indicate the need for **one nodal equation**.

FIGURE 16-22(a)   FIGURE 16-22(b)

The equations are:

$$K_1 = G_{11}V_1 \qquad \text{One Node} \qquad \textbf{EQUATION 16-4}$$

$$\text{(a)} \quad K_1 = G_{11}V_1 + G_{12}V_2 \qquad\qquad\qquad \textbf{EQUATION 16-5}$$

$$\text{Two Nodes}$$

$$\text{(b)} \quad K_2 = G_{21}V_1 + G_{22}V_2$$

118

(a) $K_1 = G_{11}V_1 + G_{12}V_2 + G_{13}V_3$ EQUATION 16-6

(b) $K_2 = G_{21}V_1 + G_{22}V_2 + G_{23}V_3$ Three Nodes

(c) $K_3 = G_{31}V_1 + G_{32}V_2 + G_{33}V_3$

where: K_1 =   the algebraic sum of all short circuit branch currents entering node 1. If the arrow leaves the positive terminal of the sources, then a positive sign is placed before the branch current. If the arrow enters the positive terminal, then a negative sign is placed before the branch current.

G_{11} =   the "mesh" conductance. This conductance is obtained by taking the reciprocal of the total resistance, of each branch, that terminates at the node under investigation. The question that is asked is: "What conductances terminating at node 1 (first subscript) are influenced by the voltage of node 1 (second subscript)?" The mesh terms ($G_{11}V_1$, $G_{22}V_2$, $G_{33}V_3$, etc.) are always positive.

V_1 =   the assumed voltage for node 1.

K_2 =   use the same explanation that describes K_1 except it is node 2 that is under investigation.

G_{12} =   the "mutual" conductance is the conductance between node 1 and node 2. The question is: "Which conductances terminated at node 1 are influenced by the voltage of node 2?" The mutual terms ($G_{12}V_2$, $G_{23}V_3$, $G_{32}V_2$, etc.) are **always negative**.

All remaining coefficients are self-explanatory.

RULES FOR SOLVING NODAL EQUATIONS ARE:

1. **The first rule in setting up a network for nodal analysis is to "declare" each and every node the most positive in the network with respect to the reference node.** Following this rule for the network of Figure 16-23, the current through the three branches, or legs, of the circuit will have to be drawn away from the reference node and toward the positive node. The current arrows that define these conditions have also been included. An examination of these arrows clearly indicates that at least one of the currents must be leaving the node. Once the voltage at the node has been determined, the actual direction of current flow can be established.

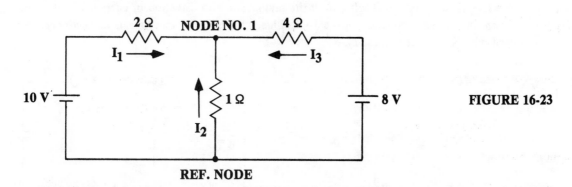

FIGURE 16-23

2. **Select one node to be investigated and write its equation.** During this investigation, every remaining node is to be shorted to the reference point.
3. **The magnitude of each branch current can be evaluated if each branch is assumed to be a Thevenin's equivalent.** Dividing the voltage of the branch by the resistance of the branch will result in the short circuit branch current.

119

EXAMPLE 5:

Determine the voltage E_{ab} for the circuit of Figure 16-24.

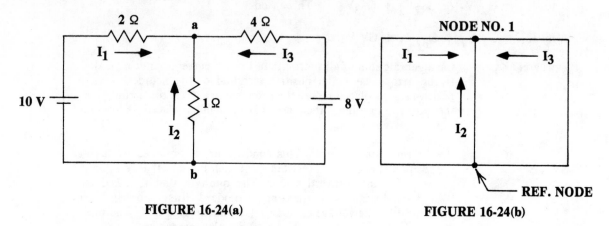

FIGURE 16-24(a)   **FIGURE 16-24(b)**

SOLUTION: Follow the rules previously outlined.

Step 1. Draw the network graph and determine the number of nodes. This graph is shown in Figure 16-24(b) with only one equation (Equation 16-4) required for the solution. Note that two loop equations would be required to obtain the same solution.

Step 2. Draw in the current arrows toward node No. 1.

Step 3. Determine K_1 using Rule 3.

$$K_1 = \frac{10\text{ V}}{2\text{ }\Omega} + \frac{0\text{ V}}{1\text{ }\Omega} + \frac{8\text{ V}}{4\text{ }\Omega} = 7\text{ A}$$

Step 4. Determine G_{11}.

$$G_{11} = \frac{1}{2\text{ }\Omega} + \frac{1}{1\text{ }\Omega} + \frac{1}{4\text{ }\Omega} = 0.5 + 1 + 0.25 = 1.75$$

Step 5. Substitute the results of Step 3 into Equation 16-4 and solve for V_1. Thus:

$$K_1 = G_{11}V_1 \text{ and } V_1 = K_1/G_{11}$$

$$V_1 = \frac{7}{1.75} = 4\text{ volts}$$

Step 6. Since E_{ab} is equal to $+4.0$ volts, both the magnitude and direction of currents I_1, I_2, and I_3, can now be determined. Figure 16-25 illustrates how the circuit can be sectioned and solved with Kirchhoff's voltage law.

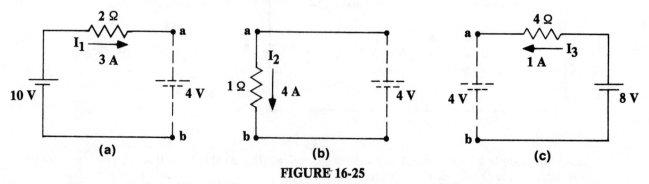

FIGURE 16-25

Circuit (a): Since the 10 V source has the highest potential, the current will flow away from the positive terminal of this source. The voltage developed across the 2-ohm resistor is $10 - 4.0 = 6.0$ V and the

120

current is

$$\frac{6.0}{2\,\Omega} = 3.0 \text{ A}$$

Circuit (b): The current flow is toward point **b** with a magnitude of

$$\frac{4.0}{1\,\Omega} = 4.0 \text{ A}$$

Circuit (c): The 8-volt source has the highest potential. The current flow is away from the positive terminal of this source with a magnitude of

$$\frac{8 - 4.0}{4} = \frac{4.0}{4} = 1.0 \text{ A}$$

Checking the results by Kirchhoff's current law:

$$I_2 = I_1 + I_3$$

$$4 \text{ A} = 3 \text{ A} + 1 \text{ A}$$

$$4 = 4$$

EXAMPLE 6:

Determine the voltages E_{ab} and E_{cb} for the circuit in Figure 16-26. In addition, determine both the magnitude and direction of the current through the 3-ohm resistor.

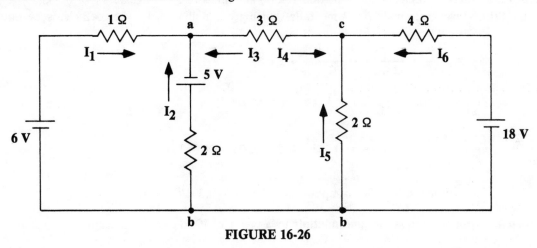

FIGURE 16-26

SOLUTION:

Step 1. Draw the network graph. See Figure 16-27.

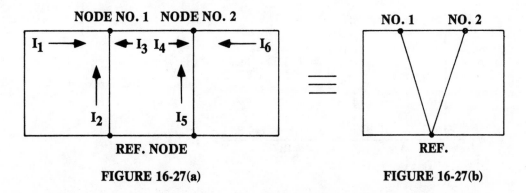

FIGURE 16-27(a)   **FIGURE 16-27(b)**

Since two nodes exist, two equations will be required. They are:

121

(a) $K_1 = G_{11}V_1 + G_{12}V_2$

(b) $K_2 = G_{21}V_1 + G_{22}V_2$

Step 2.   Declaring each node to be the most positive point in the network, draw in a set of current arrows for each node. These arrows are shown in Figure 16-27(a).

Step 3(a).   Investigate node No. 1 first. Remember to mentally short node No. 2 to the reference point. Since Equation 16-5(a) describes node No. 1, the coefficients K_1, G_{11}, AND G_{12} must be determined. Thus,

$$K_1 = \frac{6\ V}{1\ \Omega} - \frac{5\ V}{2\ \Omega} + \frac{0\ V}{3\ \Omega} = 6 - 2.5 = 3.5$$

$$G_{11} = \frac{1}{1} + \frac{1}{2} + \frac{1}{3} = 1 + 0.5 + 0.333 = 1.833$$

$$G_{12} = -\frac{1}{3} = -0.333$$

The G_{12} value was determined by asking: "Which conductances terminated at node No. 1 are influenced by the voltage of node No. 2. The only conductance that has more than one current arrow flowing through it is the conductance of the 3-ohm resistor tied between node No. 1 and node No. 2. Remember the coefficient is negative.

Substituting these coefficients into Equation 16-5(a) simplifies to $3.5 = 1.833\ V_1 - 0.333\ V_2$.

Step 3(b).   Investigate node No. 2 mentally shorting out node No. 1. The coefficients of Equation 16-5(b) are:

$$K_2 = \frac{0\ V}{3\ \Omega} + \frac{0\ V}{2\ \Omega} + \frac{18\ V}{4\ \Omega} = 4.5$$

$$G_{21} = \text{same as } G_{12} = -\frac{1}{3} = -0.333$$

$$G_{22} = \frac{1}{3} + \frac{1}{2} + \frac{1}{4} = 0.333 + 0.5 + 0.25 = 1.083$$

Substituting into 16-5(b) simplifies to: $4.5 = -0.333V_1 + 1.083V_2$

Step 4.   Group the equations and multiply by a factor of 10 to clear the decimals. Thus:

Equation 16-5(a): $35 = 18.33V_1 - 3.33V_2$

Equation 16-5(b): $45 = -3.33V_1 + 10.83V_2$

Solving by determinants:

$$\Delta = \begin{vmatrix} 18.33 & -3.33 \\ -3.33 & 10.83 \end{vmatrix} = 198.5 - 11.09 = 187.4$$

$$V_1 = \frac{\begin{vmatrix} 35 & -3.33 \\ 45 & 10.83 \end{vmatrix}}{\Delta} = \frac{379.05 + 149.85}{187.4} = 2.82\ V$$

$$V_2 = \frac{\begin{vmatrix} 18.33 & 35 \\ -3.33 & 45 \end{vmatrix}}{\Delta} = \frac{824.85 + 116.55}{187.4} = 5.02\ V$$

Step 5.   Redraw the circuit in sections as shown in Figure 16-28. Note the E_{cb} is higher in potential than E_{ab}; therefore, the current arrow is drawn toward Point **a** as shown.   The current is:

$$\frac{5.02 - 2.82}{3} = \frac{2.20}{3} = 0.733 \text{ A}$$

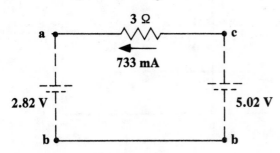

FIGURE 16-28

EXAMPLE 7 :
Determine the remaining currents of Figure 16-26.
SOLUTION: Draw in the remaining sections. See Figure 16-29.

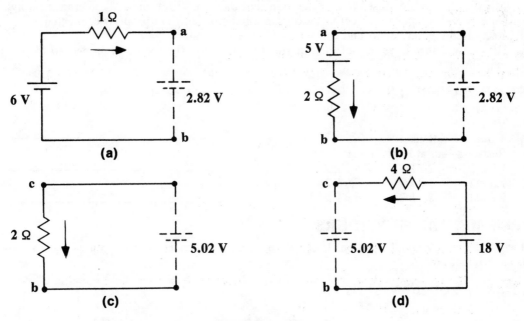

FIGURE 16-29

Circuit (a): $I_1 = \dfrac{6.0 - 2.82}{1 \ \Omega} = \dfrac{3.18}{1} = 3.18 \text{ A}$

Circuit (b): $I_2 = \dfrac{2.82 + 5}{2 \ \Omega} = \dfrac{7.82}{2} = 3.91 \text{ A}$

Circuit (c): $I_5 = \dfrac{5.02}{2 \ \Omega} = 2.51 \text{ A}$

Circuit (d): $I_6 = \dfrac{18 - 5.02}{4 \ \Omega} = \dfrac{12.98}{4} = 3.245 \text{ A}$

Redrawing the circuit and labeling the currents results in Figure 16-30.

123

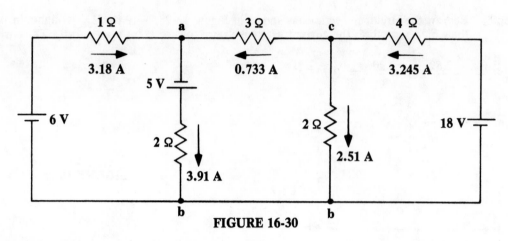

FIGURE 16-30

Using Kirchhoff's current law, the currents check.

SUGGESTED OBSERVATIONS

1. Make calculations prior to performing the experiment.
2. Measure the resistors used to make certain that they are within the 10% tolerance rating.
3. Construct the circuits with care. Be certain all components are properly connected.
4. Record all measurements and calculations in the data table.
5. Retain, and turn in as part of the experimental write-up, the computation needed to complete this experiment.

LIST OF MATERIALS   Alternate Materials
1. MFM
2. VPS
3. Three 1.5 volt cells
4. Resistors—all at least ½ watt

1 kΩ	2.7 kΩ	4.7 kΩ
1.8 kΩ	3.3 kΩ	5.6 kΩ
2.2 kΩ	3.9 kΩ	6.8 kΩ

EXPERIMENTAL PROCEDURE

You will use loop and nodal equations to calculate circuit parameters and then you will verify your calculations experimentally.

SECTION A: NODAL ANALYSIS

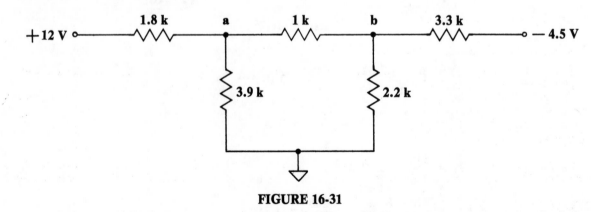

FIGURE 16-31

1. Using nodal analysis, calculate E_{ba}. Include these calculations as part of the experimental write-up.
2. Construct the circuit of Figure 16-31 and measure E_{ba}.
3. Compare the measured and calculated values of E_{ba}.

SECTION B: LOOP ANALYSIS

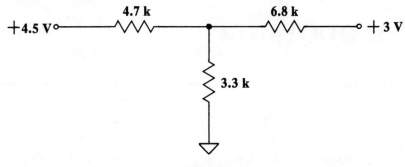

FIGURE 16-32

1. Using loop analysis, mathematically determine the current through the 3.3 kΩ resistor.
2. Construct the circuit of Figure 16-32 and experimentally determine the amount of current passing through the 3.3 kΩ resistor.
3. Compare the measured and calculated values of current.

SECTION C: LOOP ANALYSIS

1. Mathematically determine the current through the 3.3 kΩ resistor of Figure 16-33. Three loop equations are needed for the solution.
2. Construct the circuit of Figure 16-33 and measure the current through the 3.3 kΩ resistor.

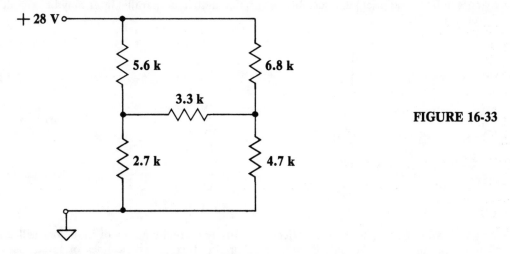

FIGURE 16-33

3. Compare the measured and calculated values of current.

DATA INTERPRETATION AND CONCLUSIONS

Write a general summary of the ideas presented in this experiment. This discussion should include:
1. The outlined procedure used to solve for voltage using nodal analysis.
2. The outlined procedure used to solve for current using loop equations.
3. Your own conclusions.

APPLICATIONS

Loop and nodal equations are powerful analytical tools for circuit analysis. Once a circuit has been analyzed, it is then possible to determine if the same answers appear in a practical laboratory situation.

PROBLEM

1. Use nodal analysis to solve for the current through the 3.3 kΩ resistor in Figure 16-32.

17 | INDUCTANCE

OBJECTIVES

With the exception of **heat-power** applications (electric lights, electric toasters, electric soldering irons, etc.), purely resistive circuits are limited in practical application. For instance, resistance cannot convert electrical energy to mechanical energy to ring a bell, close relay contacts, or rotate a motor. These loads all require coils of wire as the driving force. Inductance is the inherent property of a coil. This experiment introduces the characteristics of inductance. The effects of inductors, in various circuit configurations, are investigated experimentally.

INDUCTANCE

Inductance is defined as the property of a circuit or coil that causes a counter-electromotive force to be generated which will, in turn, cause the circuit or coil to oppose any change in the current flowing through that circuit or coil. The unit for inductance is the HENRY designed by "H" and its schematic letter symbol is "L". The three schematic symbols shown in Figure 17-1 are: (a) an air core inductor, (b) a ferromagnetic core inductor, and (c) an adjustable or variable inductor—parallel lines may be included to indicate a ferromagnetic core.

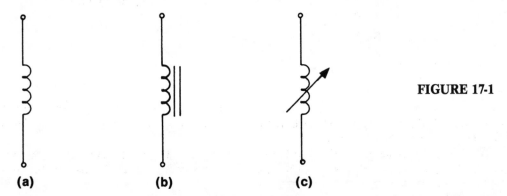

(a)   (b)   (c)

FIGURE 17-1

Equations 17-1 and 17-2 define the total effect that different combinations of inductors will make in a circuit. These equations are valid only if there is no linking flux (magnetic lines of force) between one inductor and another. Equation 17-1 is used when inductors are connected in series and Equation 17-2 is used for inductors connected in parallel.

$$L_t = L_1 + L_1 + ... + L_n \qquad \text{EQUATION 17-1}$$

$$L_t = \cfrac{1}{\cfrac{1}{L_1} + \cfrac{1}{L_2} + \cfrac{1}{L_3} + ... + \cfrac{1}{L_n}} \qquad \text{EQUATION 17-2}$$

where: L_t = total inductance offered by the combination

L_1, L_2, etc. = inductance of the individual inductors

SUGGESTED OBSERVATIONS

1. CAUTION: Inductors can produce voltages capable of causing a painful shock. It is imperative that only **ONE** hand be used when connecting and disconnecting inductors within a circuit. **All** sources

126

should be removed prior to handling inductors.

2.   Since some windings on transformers have very low values of resistance, excessive current can flow if a direct current potential is connected directly across the winding. Consult the instructor to determine what is considered to be excessive for the transformer being used.

3.   Record all measurements and observations in the data table.

LIST OF MATERIALS

Alternate Materials

1.   MFM
2.   VPS
3.   1.5 Volt Cell
4.   Two Output Transformers
5.   Headphone (2 kΩ)
6.   Three Neon Lamps NE-2
7.   4.7 Ω Resistor—½ watt
8.   150 kΩ Resistor—½ watt

EXPERIMENTAL PROCEDURE

In this experiment you will examine the effects of connecting inductors in a system and the process of coupling the flux of one unit to another. A small output transformer is used which is indicated by the schematic symbol shown in Figure 17-2. Since the transformer is simply a tool for this experiment, it will not be discussed at this time.

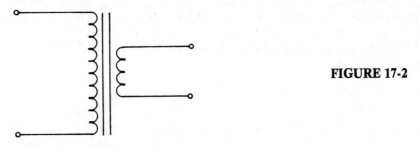

FIGURE 17-2

SECTION A: INDUCED VOLTAGE

To visually demonstrate that a collapsing magnetic field can induce a high voltage, a neon lamp is used. Since a neon lamp only lights when a predetermined voltage level is exceeded, it will provide a relative indication of the **minimum** voltage generated.

1.   Place a NE-2 neon lamp across a VPS as in Figure 17-3(a). (The 150 kΩ resistor limits the current to protect the lamp.) Adjust the VPS until the NE-2 just fires. Record this voltage (approximately 80 to 90 volts).

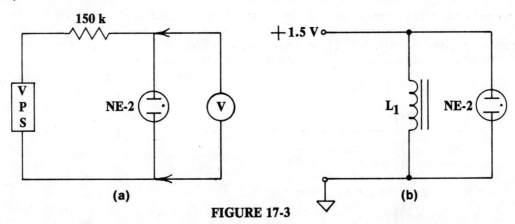

FIGURE 17-3

2.   Determine the resistance of each winding of an output transformer. Use the winding with the highest resistance (100 - 200 Ω) as L_1 and construct the circuit of Figure 17-3(b). **Leave the voltage source disconnected.** Disregard the other windings on the transformer.

3.   Demonstrate that the inductor has inductance by momentarily (less than a second) completing the

127

circuit of Figure 17-3(b); then quickly open it. The lamp should flash.

4. In a short statement, account for the apparent increase in source voltage.

SECTION B: INDUCTION (CLOSELY COUPLED)

1. Construct the circuit of Figure 17-4. Connect the NE-2 to the high resistance winding of the transformer. **Leave the source disconnected.**

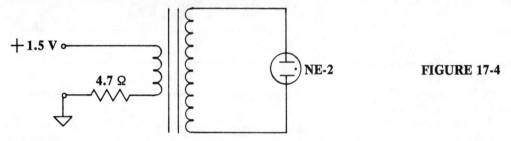

FIGURE 17-4

2. Momentarily complete the circuit (less than a second); then, quickly open it. The NE-2 should flash.
3. Account for the presence of enough voltage to fire the NE-2.

SECTION C: INDUCTION (LOOSELY COUPLED)

1. Using two output transformers, construct the circuit of Figure 17-5. Attach one terminal of the battery to the low resistance winding (0.1 to 0.5 Ω) of transformer T_1. Leave the other battery terminal disconnected. Place the headphones across the high resistance winding of output transformer T_2.

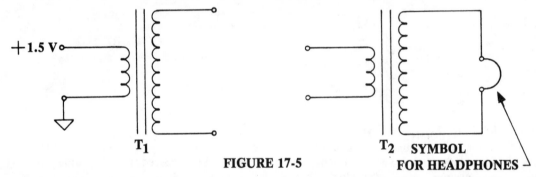

FIGURE 17-5

T_1

T_2 SYMBOL
FOR HEADPHONES

2. Place the two transformers close to one another (less than an inch apart), but not touching. While listening with the headphones, quickly (less than a second) make and break the circuit containing the battery. (Remember, never leave the battery connected as it may damage the transformer!) A noise should be heard which indicates current is flowing in the headphone circuit.
3. How did current flow through the headphone when there is no visible voltage source?

SECTION D: AIDING AND OPPOSING INDUCED VOLTAGES

1. Construct the circuit of Figure 17-6. Leave the voltage terminals disconnected.

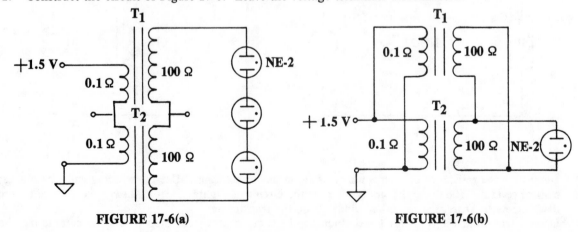

FIGURE 17-6(a)

FIGURE 17-6(b)

2. Momentarily, complete the circuit containing the battery. Note the relative brilliance of the neon lamps.
3. Reverse one transformer's secondary (100 Ω winding) leads. Again, momentarily complete the circuit. Note the relative brilliance of the neon lamps.
4. Account for the difference in brilliance.
5. Place two transformers in parallel as in Figure 17-6(b). Leave the voltage terminals disconnected.
6. Repeat Steps 2 through 4.

DATA INTERPRETATION AND CONCLUSION

Write a general summary of the ideas presented in this experiment. This discussion should include:
1. A definition of inductance and induction.
2. A description of transformer action.
3. Your own conclusions.

APPLICATIONS

There are numerous applications for the inductor, with a majority falling in the filtering category. Some of these applications are: (1) radios, (2) public address systems, (3) wireless intercommunication systems, (4) television sets, (5) transmitters, and (6) electronic recording equipment.

PROBLEM

1. A circuit requires a 0.7 Henry inductor with an approximate current rating of 200 mA. Using the industrial parts catalog, compare the prices, dimensions, and weight of this inductor as it is listed by two or three different manufacturers.

NOTE: This information may be listed under inductors, chokes, filter chokes, filter reactors, smoothing chokes or toroids.

18 | CAPACITANCE

OBJECTIVES

In addition to resistance and inductance, every electrical circuit has capacitance. In component form, the capacitor is used in electronic circuits almost as extensively as the resistor. This experiment introduces the student to capacitance and the electrical characteristics of the capacitor.

CAPACITANCE

Capacitance is the property of an electric circuit that opposes changes of voltage across the circuit. The unit for capacitance is the FARAD designated by "F" and its schematic letter symbol is "C". The three most common schematic symbols, shown in Figure 18-1, are: (a) the fixed capacitor (the curved element represents the outside or negative electrode), (b) the polarized capacitor (electrolytic), and (c) the adjustable or variable capacitor. (The arrow represents the moving element.)

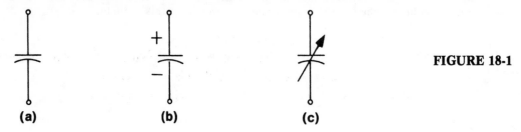

FIGURE 18-1

(a)   (b)   (c)

Essentially, capacitance is the amount of charge a capacitor can store on its plates. It is defined as the ratio of the charge stored to the voltage producing the charge. Mathematically expressed:

$$C = \frac{Q}{E}$$

EQUATION 18-1

where: C = the designated symbol for capacitance is expressed in its basic unit: the farad
Q = the amount of charge (in coulombs) held in a dielectric
E = the impressed voltage

One feature of the capacitor is that its charge cannot be changed instantaneously. Therefore, since C and Q are constant at any instance, Equation 18-1 is used to show that the voltage across the capacitor cannot be changed instantaneously.

The following equations define capacitor action in series and parallel configurations. For the series configuration, Equation 18-2 states that the charge is the same in each capacitor.

$$Q_T = Q_1 = Q_2 = ... = Q_n$$

EQUATION 18-2

where: Q_T = the total charge in the series configuration

$Q_1, Q_2 ... Q_n$ = individual charge in each capacitor

Equation 18-3 states the total capacity for two or more capacitors, connected in series, is always less than the value of the lowest rated capacitance.

$$C_T = \frac{1}{\dfrac{1}{C_1} + \dfrac{1}{C_2} + ... + \dfrac{1}{C_n}}$$

EQUATION 18-3

Equation Descriptions: where: C_T = the total capacitance of the series configuration

C_1, C_2,... C_n = capacitance of the individual capacitors

For the parallel configuration, Equation 18-4 states that the total capacitance is the sum of individual capacitances, while Equation 18-5 defines the total charge as being the sum of the charges contained within each capacitor.

$$C_T = C_1 + C_2 + ... + C_n$$ **EQUATION 18-4**

$$Q_T = Q_1 + Q_2 + ... + Q_n$$ **EQUATION 18-5**

Manufacturers specify the capacitance value of their capacitors and the maximum voltage that can be continuously impressed across the capacitor. It is referred to as the "working volts DC". Under NO circumstances must this voltage rating be exceeded.

Electrolytic capacitors require additional considerations. In many electrolytics, the rated capacity is maintained by operating the capacitor with at least 80 % of its rated voltage. If, for example, a capacitor rated at 100 μF at 450 V is operated at 275 V for any length of time, the capacity will drop in value.

Leakage currents in a properly operating electrolytic capacitor range from microamperes to milliamperes, depending upon the type of capacitor.

The equivalent circuit for a capacitor is shown in Figure 18-2.

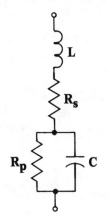

where: C = the capacitance of the capacitor

R_s = resistance of the leads, lead connections and plate material

R_p = resistance due to the dielectric material

L = inductance of the leads and the capacitor plates

FIGURE 18-2

Equation 18-6 approximates the maximum permissible leakage current for commercial electrolytic capacitors using aluminum etch foil plates.

$$I(mA) = KC(\mu F) + 0.3$$ **EQUATION 18-6**

where: K = 0.01 for   3 - 100 V
        0.02   for 101 - 150 V
        0.025 for 251 - 350 V
        0.04   for 351 - 450 V

Equation 18-7 approximates the maximum allowable leakage current for a tantalum electrolytic capacitor.

$$I(\mu A) = 0.003C(\mu F) = 0.003C(\mu F)V$$ **EQUATION 18-7**

EXAMPLE 1:

What is the maximum allowable leakage current for a 16 μF aluminum foil electrolytic capacitor rated at 450 volts?

SOLUTION: Using Equation 18-6, solve the problem.

$$I = 0.04 \times 16 + 0.3 = 0.64 + 0.3 = 0.94 \text{ mA}$$

131

EXAMPLE 2:

What is the maximum allowable leakage current for a 10 μF tantalum capacitor rated at 100 volts?
SOLUTION: Solve by Equation 18-7. **See Figure 18-2.**

$$I = 0.003 \times 10 \times 100 = 3 \ \mu A$$

SUGGESTED OBSERVATIONS

1. Discharge capacitors before and after use. Capacitors are discharged by shorting the leads or terminals.
2. Record all readings, observations, and calculations in the data table.

LIST OF MATERIALS

Alternate Materials

1. MFM
2. VOM
3. VPS
4. Capacitors—at least 100 VDC
    0.1 μF   1 μF (2)
    0.5 μF   100 μF (polarized)
5. Two 100 kΩ Resistors

_____

_____

_____

_____

EXPERIMENTAL PROCEDURE

In this experiment you will become acquainted with the process of redistribution of charges in capacitors both for the series and parallel configurations.

SECTION A: CAPACITIVE CIRCUIT - PARALLEL

1. Calculate the charge stored in a 1 μF capacitor that has been charged with a 100-volt source.
2. Charge a 1 μF capacitor to 100 V. See Figure 18-3. Carefully remove the charged capacitor (one lead at a time) from the charging circuit. **(Remove the meter from the circuit prior to removing the capacitor.)**

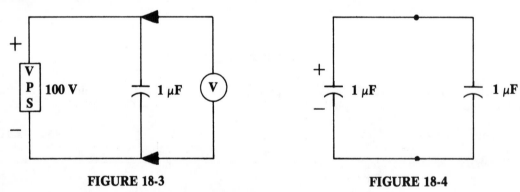

FIGURE 18-3   FIGURE 18-4

3. Place the charged capacitor in parallel with another uncharged 1 μF capacitor as in Figure 18-4. Using an analog MFM, measure the voltage across the capacitors.
4. Using the known quantity of charge (Step 1) where Q = CE and Equation 18-1, determine the voltage across the capacitors of Figure 18-4. C = 2 μF.
5. Compare the calculated voltage to the measured voltage (Steps 3 and 4).
6. Using the calculated equivalent capacitance of Figure 18-4 and the voltage of Step 3, calculate the quantity of charge contained in this circuit.
7. Compare the charge determined in Step 6 to that of Step 1.

SECTION B: CAPACITIVE CIRCUIT - SERIES

1. **Be sure to discharge capacitors before use.** Construct the circuit of Figure 18-5. Charge the capacitors. Carefully remove them from the circuit. Quickly, measure the voltage across each capacitor. (If this step is repeated, discharge each capacitor before placing it back into the circuit.)
2. Using Equation 18-1, determine the charge contained within each capacitor.
3. Determine the total charge. Use the applied voltage and the total capacitance determined by Equations 18-3.

4. Compare each value of charge determined in Step 2 to the total charge determined in Step 3.

SECTION C: EQUALIZING VOLTAGE ACROSS SERIES CAPACITORS

   To prevent uneven voltage distribution across series connected capacitors, a circuit like that of Figure 18-6 is used.

1. Connect two 1 μF capacitors in series and apply 100 volts. Quickly measure the voltage across each capacitor. (If the meter is left a few seconds, a decay of voltage will be noted. While the voltage is decaying, due to the meter loading, the other capacitor's voltage is rising.) Are the measured voltages equal?
2. Construct the circuit of Figure 18-6. Once again measure the voltage across each capacitor. Is the voltage stable?

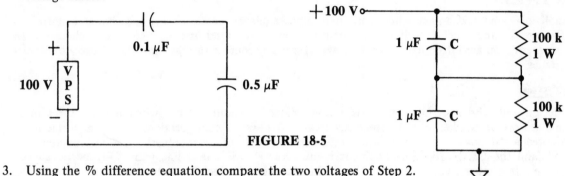

FIGURE 18-5

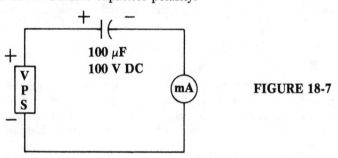

FIGURE 18-6

3. Using the % difference equation, compare the two voltages of Step 2.

SECTION D: CAPACITOR LEAKAGE DETERMINATION
1. Construct the circuit of Figure 18-7. Observe capacitor polarity.

FIGURE 18-7

2. Watch the milliammeter while adjusting the VPS to 40 volts. (If a current greater than 3 mA is noted, then the capacitor needs forming. Consult the instructor.)
3. Once the current has stabilized, record the milliammeter reading.
4. Using the source voltage and the circuit current, determine the equivalent resistance of the capacitor. (R_p = E/I). This is R_p of Figure 18-2.
5. Draw the parallel equivalent circuit of the capacitor of Figure 18-7.

DATA INTERPRETATION AND CONCLUSIONS

   Write a general summary of the ideas presented in this experiment. This discussion should include:
1. The distribution of charge in a parallel capacitive circuit.
2. The distribution of voltage in a series capacitive circuit.
3. Your own conclusions.

APPLICATIONS

   Capacitors, like resistors and inductors, are used in every conceivable piece of electronic equipment. Several applications include: (1) filter circuits, (2) coupling circuits, and (3) timing circuits.

PROBLEMS

1. Why were 100 kΩ resistors used in the circuit of Figure 18-6?
2. What wattage is each resistor dissipating in Figure 18-6?
3. Determine the cost of a 20 μF capacitor rated at appoximately (a) 12 volts, (b) 25 volts, and (c) 50 volts.

19 | SWITCHES — CHARACTERISTICS AND RATINGS

OBJECTIVES

Manually operated mechanical switches are used to make (close), break (open), and change connections in electric circuits and systems. Switches represent the most basic kind of circuit control mechanism. In this experiment, you are introduced to contact arrangements, contact ratings, and the control of circuits with switches.

SWITCHES

The design and fabrication of a switch is a complicated procedure. The switch manufacturer must produce a variety of switches with different electrical and physical configurations. Most manufacturers have adopted a "standard line" which accommodates the needs of the hobbiest, student, maintenance personnel, and the OEM (original equipment manufacturer). Standard line items may be purchased from local electronic parts supply stores. The OEM may need a special switch. When this need arises, the OEM submits his specifications to the switch manufacturer who, in turn, designs and fabricates the switch.

Since the primary function of a switch is to control circuitry, it must be selected on the basis of contact arrangement and contact rating. The physical size of the switch, while an influencing factor, cannot always be a basis of selection since contact arrangements and ratings determine the minimum physical size.

CONTACT ARRANGEMENTS

Contact arrangement is defined in terms of "poles" and the number of active positions in which the poles can be placed. The symbol shown in Figure 19-1 is used to indicate that one wire can be controlled by either opening or closing the circuit. An illustration of the usage of this type of switch is shown in Figure 19-2 This is the schematic of a simple flashlight. Since there is only one controlling, or moving,

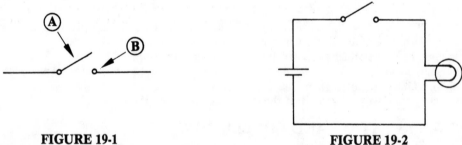

| FIGURE 19-1 | FIGURE 19-2 |

element (Point A of Figure 19-1), the switch has one "pole". The only active position that the pole can be placed in is the contact indicated as Point B. Thus, the switch is identified as a "single pole single throw" switch. This is abbreviated to SPST. Additionally, since the switch is shown in the open position, it is classed as "normally open" (N.O.). Figure 19-3 shows the symbol for a SPST N.C. ("normally closed") switch. A single pole double throw (SPDT) switch is shown in Figure 19-4.

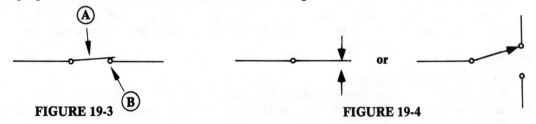

| FIGURE 19-3 | FIGURE 19-4 |

134

The switches shown in Figure 19-5 are: (a) and (b) double throw (DPDT), and (c) and (d) three pole double throw (3PDT).

NOTE: The dashed lines indicate that the poles are connected together in a manner which allows all poles to move simultaneously from one position to another.

(a)

(b)

FIGURE 19-5

(c)

(d)

The rotary switch is described in terms of "poles and positions". A single pole five position switch symbol is shown in Figure 19-6. An additional specification, included for rotary and lever switches, is the manner in which the contacts make and break. For the rotary switch there is the non-shorting (non-bridging), break-before-make arrangement, see Figure 19-6, and the shorting (bridging), make-before-break arrangement shown in Figure 19-7. Contact arrangements for lever switches and relays are the same. A few are shown in Table 19-1.

FIGURE 19-6   **FIGURE 19-7**

Draftsmen sometimes include a physical diagram of a multiple-multipurpose rotary switch. A three pole four position wafer of a rotating switch is shown in Figure 19-8. Observe that it contains two bridging sections and one non-bridging section.

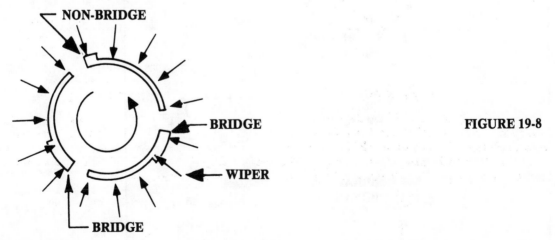

FIGURE 19-8

CONTACT RATING

The contact rating of a switch includes: (1) the maximum value of voltage that can be applied to the contacts (exceeding this voltage rating may cause a flash-over to other contacts. to ground, or both);

(2) the maximum current that can be switched, the maximum voltage level for the maximum current specified, and the type of current applied (alternating or direct current); and (3) whether the load is resistive or inductive. Examples 1 and 2 illustrate how manufacturers' specifications are interpreted.

EXAMPLE 1:
General Purpose Switch
Contact arrangement: SPDT
Ratings: 15 amps,   125, 250, or 480 volts AC
        ½ amp,   125 volts DC
        ¼ amp,   250 volts DC
        ¼ HP,   125 volts AC
        ½ HP,   250 volts AC

SOLUTION:

General Purpose—this switch is economically priced and is used where precision is not required.

FORM	TERM	CONTACT CONFIGURATIONS
A	MAKE	
B	BREAK	
C	BREAK MAKE (transfer)	
D	MAKE - BEFORE BREAK (continuity transfer)	
E	BREAK MAKE - BEFORE BREAK	
F	MAKE MAKE	
G	BREAK BREAK	
H	BREAK BREAK MAKE	

TABLE 19-1

136

Contact arrangement—form C—single pole double throw (SPDT).

Ratings—this switch can open or close a circuit that is carrying a maximum current of 15 amperes at any of the alternating current voltages specified. (a) If direct current voltage is used, the maximum current that can be switched is ½ ampere at 125 volts DC or ¼ ampere at 250 volts DC. (The reason for the reduction in maximum current for DC circuits is that DC causes the switch contacts to burn and pit.) (b) When the switch is used to turn motors on or off, the switch is rated at the maximum horsepower that can be handled. Thus, a ¼ HP motor that is going to operate on a 125 volt AC line or a ½ HP motor operating on a 250 volt AC line is the maximum horsepower capacity that this switch can safely handle. (c) Motors require a starting current that is approximately 3 to 5 times the running current. Since 1 HP is 746 watts, a ¼ HP motor operating at 125 volts has a running current of approximately

$$\frac{186.5 \text{ watts}}{125 \text{ volts}} = 1.5 \text{ A.}$$

The startng current could exceed 7.5 amperes. Also, the load is considered to be inductive and the current-voltage combination must be carefully controlled to reduce the burning and pitting of the contacts.

EXAMPLE 2

Toggle Switch—General Purpose
Contact Arrangement: 3PDT
Electrical Ratings:
   Resistive: 4 amp,   30 V DC or 125/250 V AC
   Inductive: 2.5 amp,   30 V DC
   Lamp: 2.4 amp,   30 V DC or 125/250 V AC
Contact Material: Fine Silver

SOLUTION: In addition to the call-outs of Example 1, the ratings in terms of resistive-inductive and lamp usage is specified. Lamp usage was the only topic not covered in Example 1. Recalling that the static cold resistance of a lamp is from 1/6 to 1/10 the hot resistance, it can be seen that the **inrush** current through the cold resistance will be 6 to 10 times greater than the operating current. Therefore, the inrush current must be considered when specifying switch parameters. Inrush current can be computed if the maximum operating current of the lamp is known.

SUGGESTED OBSERVATIONS

1.   To minimize wiring errors, exercise care when constructing the circuits of this experiment.
2.   The various switch configurations (SPST, SPDT, etc.) are obtained from a DPDT switch.
3.   Place all measurements, calculations, and schematics in the data tables.

LIST OF MATERIALS   Alternate Materials

1.   Four 1.5 Volt Cells
2.   Four DPDT Switches
3.   Two No. 47 Lamps

EXPERIMENTAL PROCEDURE

In this experiment, contact arrangements and switch applications will be investigated.

SECTION A: SWITCHING CIRCUITS

1.   Connect the circuit of Figure 19-9 so the lamp lights when the switch is closed.

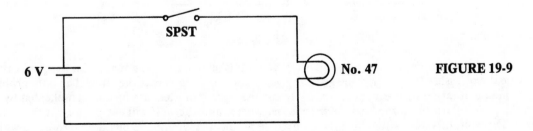

FIGURE 19-9

2 Using the schematic of Figure 19-10 construct the circuit so each switch will control the light.

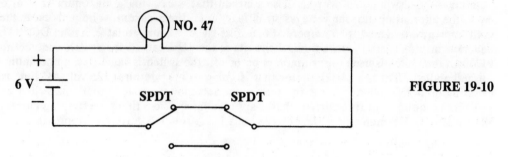

FIGURE 19-10

3. Using Figure 19-11, wire the circuit so each switch controls the light.

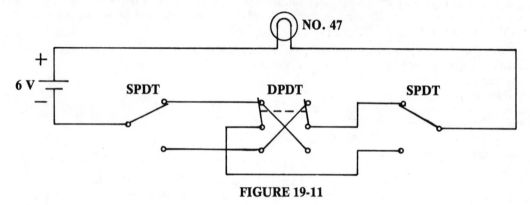

FIGURE 19-11

4. The circuit of Figure 19-12, when correctly wired, can be made into a logic game.

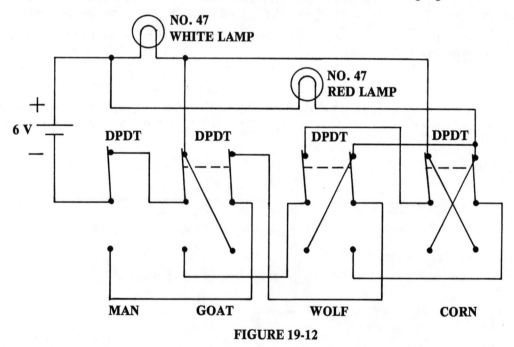

FIGURE 19-12

The four switches represent a man, a goat, a wolf, and a bushel of corn. The man, who lives on a river, has a row boat and wishes to transport the goat, the wolf, and the corn across the river. The only problem is that the man can take only one item at a time across the river. Matters are further complicated by the fact that the goat will eat the corn when the man is not there, and the wolf will eat the goat if the man is not there. The wolf, however, will not eat the corn. The problem, therefore, is one of determining which, and in what order, to transport the items across the river.

138

As long as the white light is on, the game continues, but when the red light comes on, something has been consumed. The game is started with all switches in a down position. The switch representing the man and any other switch are simultaneously moved to the up position. This move represents the man taking something across the river. The item taken across when the man is left (switch in up position) and the "man switch" is returned to the down position. Items are then conveyed back and forth across the river in a definite sequence. The game is won when all switches are in the up position and the red light has not come on. If the red light comes on, all switches are returned to the down position and the next player takes his turn.

SECTION B: DESIGN PROBLEM

Using the following materials, design and construct a circuit that can control one light from four places. You will need two SPDT switches, two DPDT switches, a No. 47 lamp, and a 6 V source. Demonstrate to the instructor that the light can be controlled from each switch. Have the instructor initial the data table. Draw the schematic of the circuit in the space provided in the data table.

DATA INTERPRETATION AND CONCLUSIONS

Write a general summary of the ideas presented in this experiment. This discussion should include:
1. The need for an understanding of schematic diagrams.
2. Switch ratings.
3. Contact arrangements.
4. Your own conclusions.

APPLICATIONS

The switch finds application in practically every electronic and electrical device—this includes the field of automotive electricity and household appliances.

PROBLEM

1. Use the ideas about the diode presented in Experiment 14 and the material covered in this experiment to design a circuit that will solve the following problem.

You are presented with two "black boxes". Upon inspection (see Figure 19-13), it can be seen that the boxes are connected with two wires. Furthermore, Box A has a switch handle visible while Box B has two lamps visible. The lamps are No. 47 bulbs. A 6 volt source capable of operating these lamps is located in Box A.

When the swtich is in one position one light lights, and when it is moved to the other position, the other light lights. However, only one light is on at a time. Using schematic symbols, draw the circuitry needed to control the lights.

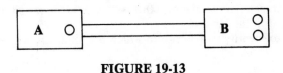

FIGURE 19-13

20 | RELAY CHARACTERISTICS

OBJECTIVES

Relays are electrical switching devices used to control electric circuits and systems. Since relay coils can operate on low power to control circuits of high power, they are found in practically all types of industrial electronic systems. This experiment introduces relay characteristics.

RELAYS

Relays are grouped by their mode of operation. Within each group are sub-groups based on contact arrangements, contact rating, size, and method or style of enclosure. Regardless of the varieties that exist, all relays have similar characteristics.

Referring back to the contact arrangements shown in Table 19-1, it can be shown that these are the basic configurations from which the manufacturer will build the relay. As an illustration, a relay can be identified as having a 1C contact arrangement—which is another way of saying SPDT. Or, on the other hand, a relay may have several contact combinations, such as 1C-3B or 2A-6C-1B. Interpretation of the 1C-3B group identifies the relay as having one set of form C contacts and three sets of form B contacts.

The method of specifying the contact rating of a relay is the same as that used for switches. However, because of the possibility that the relay may be required to cycle (turn on and off) repeatedly, some form of contact protection may be specified by the manufacturer. If such a specification exists, do not ignore it, since severe contact damage will usually occur.

The current required to cause the relay to close, or operate, is also called the **pull-in**, or **pick-up** current. The maximum value of current that will cause the relay to turn to the de-energized condition is called the **drop-out** current. Thus, a relay may consistently close at 9 mA and drop-out at 3 mA. The pull-in or pick-up current is 9 mA, while the maximum current for the relay to drop out is 3 mA. In other words, once the relay has actuated, it will remain actuated as long as a current greater than 3 mA exists.

Another way of specifying operating requirements is in terms of wattage. A relay having a maximum "coil dissipation" of 3 watts operating at 6 volts has a pull-in current of 500 mA. With this type of rating, neither the pull-in or drop-out current is specified; however, such conditions can be determined in a laboratory with considerable ease.

Relays are designed to operate on either AC or DC voltages. If a DC relay is operated with AC, it will "chatter" and the contacts will pit and burn. An AC relay can be operated on DC if the DC current is reduced to a level comparable to, or less than, the AC current required to actuate the relay. This condition must be determined in the laboratory.

Several symbols are used to identify relays, two of which are shown in Figure 20-1.

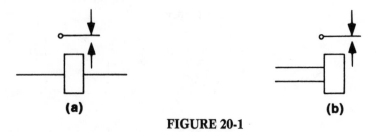

(a)   **(b)**

FIGURE 20-1

SUGGESTED OBSERVATIONS

1. Handle components with care as they are easily damaged.
2. Use the percent error equation (Equation 4-7) to compare the solutions.
3. Record all measurements, observations, and calculations in the data tables.

LIST OF MATERIALS

1. MFM
2. VOM
3. Two 22.5—45 Volt "B" Batteries
4. Four 1.5 Volt Batteries
5. Two Relays LM11 (10 kΩ)
6. Lamp No. 47 with Holder
7. Lamp No. 1819 with Holder
8. Photoconductive Cell VT34Z
9. Thermistor 41D2
10. 1 kΩ Pot
11. Two SPST Switches
12. Two Diodes 1N4001
13. Resistor—at least ½ watt:
    2.2 kΩ   10 kΩ   15 kΩ

Altenate Materials

Two Regulated VPS

_____

_____

_____

_____

_____

_____

EXPERIMENTAL PROCEDURE

In this experiment the student will examine the various relay parameters including the technique used to determine pull-in and drop-out currents.

SECTION A: RELAY CHARACTERISTICS

Two factors which are fundamental to relay operations are the pull-in and the drop-out parameters.

1. Construct the circuit of Figure 20-2.

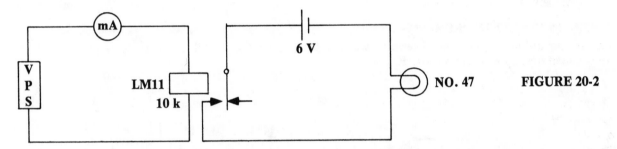

FIGURE 20-2

2. Adjust the VPS until the relay pulls in (lamp turned on). Record the current.
3. Reduce the VPS voltage until the relay drops out (lamp turned off). Record the current.
4. Repeat Steps 2 and 3 three more times and determine the average pull-in and drop-out currents.
5. Assuming a coil resistance of 10 kΩ, mathematically determine the voltage drop across the relay to cause pull-in. Use the current determined in Step 2.
6. Repeat Step 5 for the drop-out voltage. Use the current determined in Step 3.
7. Place a voltmeter across the relay and determine the pull-in and drop-out voltages.
8. Compare the voltages of Step 7 to those determined in Steps 5 and 6.

SECTION B: RELAY APPLICATION

1. Using the information about the relay obtained in Section A, mathematically determine how much voltage will be across the VT34Z when the relay in Figure 20-3 pulls in. What will the resistance of the photoconductive cell be at this point?

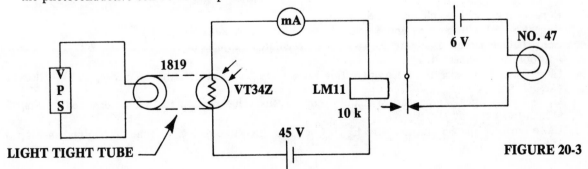

FIGURE 20-3

141

2. Construct the circuit of Figure 20-3 and monitor the voltage across the CL704L as the VPS is slowly adjusted. Record the voltage drop when the relay pulls in. Also record the pull-in current.
3. Compare the cell's calculated voltage drop (Step 1) to the measured voltage drop (Step 2).
4. Compare the pull-in current of the circuit of Figure 20-3 to that of the circuit of Figure 20-2. (Step 2, Section A and B)
5. Construct the circuit of Figure 20-4. Since the source voltage is critical, use a regulated supply or two appropriately connected 22.5 - 45 volt "B" batteries. Set the pot for the minimum resistance and leave the switch open.

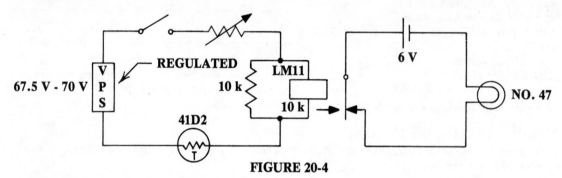

FIGURE 20-4

6. Protect the thermister from air movement. Close the switch and measure the amount of time delay before the relay closes.
7. Open the switch and allow the thermistor to cool (3 to 5 minutes). Adjust the pot for maximum resistance and repeat Step 6.

SECTION C: RELAY PROBLEM

1. Using Millman's theorem, determine which relay (K_1 or K_2) will close when S_1 of Figure 20-5 is closed and S_2 is open. Record E_{ag}.

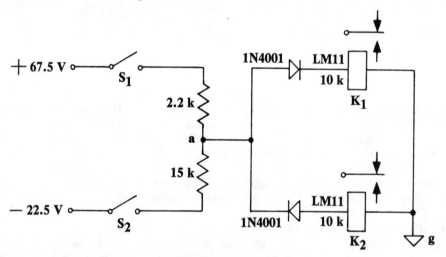

FIGURE 20-5

2. Repeat Step 1 with S_2 closed and S_1 open.
3. Repeat Step 1 with both S_1 and S_2 closed.
4. Construct the circuit of Figure 20-5. Use the regulated supplies or two 22.5 - 45 volt "B" batteries appropriately connected.
5. Close S_1 and measure E_{ag}. Compare this value of E_{ag} to that determined in Step 3. Which relay closed?
6. Close S_2 (S_1 open) and measure E_{ag}. Compare this value of E_{ag} to that determined in Step 3. Which relay closed?
7. Close both S_1 and S_2 and measure E_{ag}. Compare this value of E_{ag} to that determined in Step 3. Which relay closed?

DATA INTERPRETATION AND CONCLUSIONS

Write a general summary of the ideas presented in this experiment. This discussion should include:

1. The significance of pull-in and drop-out currents and voltages.
2. Contact arrangements.
3. Types and enclosures—see parts catalog.
4. Contact ratings.
5. Your own conclusions.

APPLICATIONS

Relays are used when only a small amount of power is available and it is necessary to control a large amount of power.

PROBLEMS

1. In the figure below, determine the value of source voltage needed to cause the relay to pull in. The relay has an 8 kΩ coil resistance requiring 5 mA for pull-in.

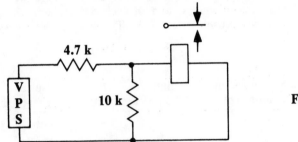

FIGURE 20-6

2. List several uses for a circuit similar to that of Figure 20-3.
3. Determine the cost of the relay used in this experiment (include the specifications).
4. List at least three different groups of relays—consult a parts catalog.

21 || RC TIME CONSTANTS

OJECTIVES

When resistance and capacitance are used in a direct current series circuit, the voltage does not rise and fall instantaneously when the circuit is energized or de-energized. It rises and falls **exponentially** with time. The rise and fall times are calculated in terms of RC time constants. By using definite values of R and C, voltage changes occur precisely and predictably. Therefore, RC timing circuits find use in motor control circuits, time control relays, and in such devices as resistance welders and radar transmitting units where high current pulses are needed. This experiment shows you how to design and apply simple RC time circuits.

RC TIME CONSTANTS

A time constant of a capacitive circuit is defined as the time required, in seconds, for the current or voltage to change by an amount equal to 63.2 percent of the total charge that will occur.

EXAMPLE 1:

In a particular circuit, a DC potential of 100 volts is placed across a series resistance-capacitance network. What is the voltage change across the capacitor after one time constant?

SOLUTION: By definition, the voltage across this element will have a voltage equal to 63.2 percent of what it previously had across it. Figure 21-1(a) shows the circuit conditions prior to applying the potential. Figure 21-1(b) shows the voltage distribuiton the instant the potential is applied. Figure 21-1(c) shows the voltage distribution after one time constant has elapsed. Finally, Figure 21-1(d) shows how the voltage is distributed after approximately five time constants have elapsed.

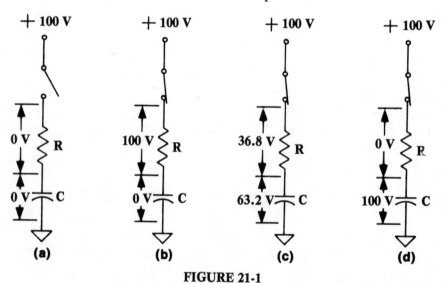

FIGURE 21-1

One of the characteristics of a capacitor is that the state of charge that exists within the capacitor cannot be changed instantaneously. Thus, in Figure 21-1(b), the state of charge in the capacitor prior to closing the switch is zero—hence zero voltage. Since the voltage across it cannot change instantaneously, the capacitor will look like a short circuit at the exact instant the switch is closed. This important concept is used in the design of filter circuits for power supplies.

EXAMPLE 2:

If the capacitor has an initial charge of 30 volts, what voltage will appear across the capacitor in Figure 21-2(a) one time constant after closing the switch?

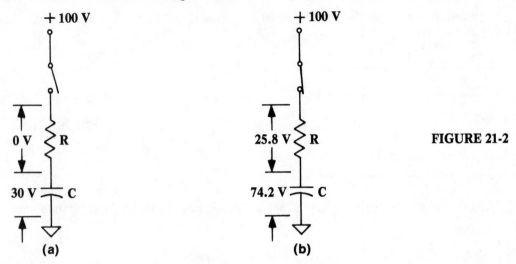

FIGURE 21-2

SOLUTION: Since a 30 volt charge exists initially and the source is 100 V, the **total** change that can occur is 70 volts. Therefore, 63.2 % of 70 volts is 44.2 volts. The voltage will be distributed as shown in Figure 21-2(b).

The circuit shown in Figure 21-3 is the **basic** or fundamental network. When working with RC time constants, the network under examination must be reduced to the basic network shown, if it does not already exist in that form.

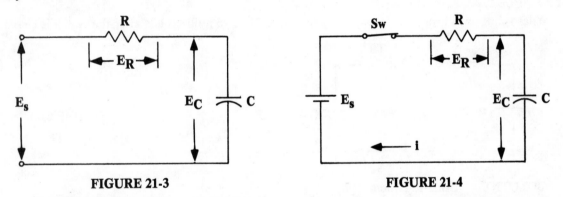

FIGURE 21-3   **FIGURE 21-4**

Four equations define the circuit parameters for the RC network shown in Figure 21-4; this circuit shows the capacitor being charged.

$$E_s = E_R + E_C$$

EQUATION 21-1

where: E_s = the voltage of the source

E_R = the voltage across the resistor

E_C = the voltage across the capacitor

$$i = \frac{E_s}{R} e^{-t/RC}$$

EQUATION 21-2

where i = the instantaneous current (A) at any given time (t)

R = circuit resistance in ohms

C = the capacitance in farads

e = constant equal to 2.718

RC = time constant of the RC circuit in seconds

t = any given time after the switch is opened or closed

145

$$E_R = E_s e^{-t/RC} \qquad\qquad\qquad\text{EQUATION 21-3}$$

$$E_C = E_s(1 - e^{-t/RC}) \qquad\qquad\text{EQUATION 21-4}$$

NOTE: Equations 21-2 through 21-4 can be solved directly with an electronic calculator. Consult your owner's guide as necessary.

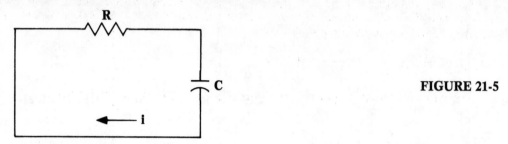

FIGURE 21-5

For the circuit of Figure 21-5, the equations that will define how the capacitor **discharges** are:

$$i = \frac{E_C}{R} e^{-t/RC} \qquad\qquad\qquad\text{EQUATION 21-5}$$

$$E_C = E_R \qquad \textbf{because C and R are in parallel} \qquad\qquad\text{EQUATION 21-6}$$

$$E_R = E_C e^{-t/RC} \qquad\qquad\qquad\text{EQUATION 21-7}$$

EXAMPLE 3:

What voltage will be across the capacitor in Figure 21-6, 75 milliseconds after the switch is closed?

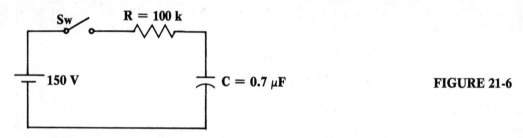

FIGURE 21-6

SOLUTION: Using Equation 21-4:

$$E_C = E_s(1 - e^{-t/RC}) = 150\left(1 - e^{-\frac{75\ ms}{70\ ms}}\right) = 150(1 - 0.343) = 150 \times 0.657 = 98.6\ V$$

EXAMPLE 4:

Determine how long it will take for the relay in Figure 21-7 to operate once the switch is closed.

**RELAY
SPECIFICATIONS**

Pull-in current = 1.8 mA
Drop-out current = 0.3 mA
Coil resistance = 8,000 ohms

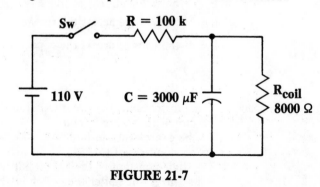

FIGURE 21-7

146

SOLUTION:

Step 1.   Detemine the voltage required to operate the relay.

$$0.8 \text{ mA} \times 8,000 \text{ }\Omega = 6.4 \text{ V}$$

Step 2.   Mentally remove the capacitor and, by using the voltage divider equation (Equation 4-5), determine if sufficient voltage will be developed across the resistance of the coil to actuate the relay. Therefore, $E_{coil} \geq 6.4$ V and by computation:

$$E_{coil} = \frac{110 \times 8 \text{ k}\Omega}{108 \text{ k}\Omega} = \frac{880}{108} = 8.15 \text{ V}$$

Step 3.   Thevenize the circuit to put it in the standard form. See Figure 21-8.

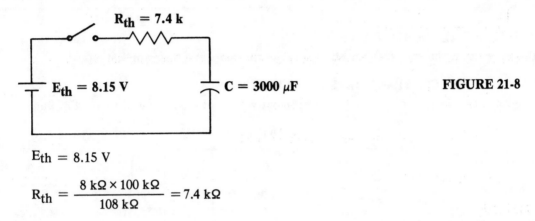

FIGURE 21-8

$E_{th} = 8.15$ V

$$R_{th} = \frac{8 \text{ k}\Omega \times 100 \text{ k}\Omega}{108 \text{ k}\Omega} = 7.4 \text{ k}\Omega$$

Step 4.   Since the capacitor is in parallel with the resistance of the relay, then the voltage required to close the relay must be the same voltage that will appear across the capacitor. Thus, t can be determined by Equation 21-4.

$$RC = 3 \times 10^{-3} \times 7.4 \times 10^{3} = 22.2 \text{ seconds}$$

$$E_C = E_s(1 - e^{-t/RC}) \qquad \text{(Equation 21-4)}$$

$$6.4 = 8.15 - 8.15e^{-t/22.2}$$

$$6.4 - 8.15 = -8.15e^{-t/22.2}$$

$$1.75/8.15 = e^{-t/22.2}$$

$$0.215 = e^{-x} \qquad \text{(where } x = t/22.2 \text{ seconds)}$$

$$\ln 0.215 = -x \ln e$$

$$x = \ln 0.215 = -1.54$$

$$x = 1.54 \text{ (drop the negative sign)}$$

$$x = t/22.2 \text{ sec.} \qquad \text{and} \qquad t = 22.2x = (22.2)(1.54)$$

$$x = 34.2 \text{ seconds}$$

SUGGESTED OBSERVATIONS

1.   Before starting the experiment, check the capacitors for excessive leakage current. If the current is excessive, the capacitor may need forming.

2.   Discharge the capacitors before handling.

3.   Record all measurements, calculations, and observations in the data tables.

LIST OF MATERIALS

1. VPS
2. VOM
3. Four 1.5 Volt Batteries
4. Two LM11 Relays 10 kΩ
5. Capacitors—at least 100 V DC
   1 μF   10 μF (2)   100 μF (3)
6. Resistors—all at least ½ watt
   4.7 kΩ (2)   10 kΩ (2)   1 MΩ
   6.8 kΩ (2)   33 kΩ
7. 1 MΩ Potentiometer
8. NE-2 Neon Lamp
9. No. 47 Lamp with Holder
10. SPST Switch

Alternate Materials
Two 45 Volt "B" Batteries

EXPERIMENTAL PROCEDURE

In this experiment you will study the effects of time constants and their applications.

SECTION A: TIME CONSTANT DETERMINATION

1. Experimentally determine the pull-in and drop-out voltage of the relay in the circuit of Figure 21-9.

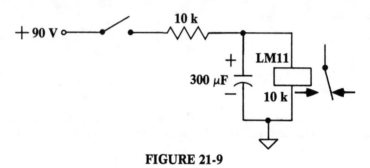

FIGURE 21-9

2. Determine the leakage current of the 300 μF capacitor (constructed from three parallel 100 μF capacitors) when 45 volts is placed across it. (See Experiment 18, Section D.) Determine the capacitor's equivalent parallel resistance.
3. Calculate the pull-in time for the relay. Use Equation 21-4 and the procedure of Example 4. Include the equivalent resistance, determined in Step 2, in the calculation.
4. Construct the circuit of Figure 21-9. Observe capacitor polarity. Verify the time calculation of Step 3. Timing should start the instant the switch is closed. After each pull-in time is determined, open the switch and carefully discharge the capacitor by momentarily shorting across it with a jumper.

NOTE: Since the timing process is only a good approximation (because of component tolerance, timing accuracy, etc.) several trials should be made.

5. Determine the percent difference between the calculated and measured values of pull-in time.
6. Use Equation 21-3 and calculate the drop-out time of the relay (Figure 21-9). In the equation, let E_R equal the drop-out voltage determined in Step 1 and let E_S equal the IR drop across the relay coil. (This is the maximum voltage the capacitor can rise to.)
7. Close the switch of Figure 21-9 and let the capacitor charge to the maximum voltage. Allow the time equal to 5 time constants for charging. Open the switch and determine the drop-out time. (Several trials should be made.)
8. Compare the computed and measured drop-out times.

SECTION B: RC TIME CONSTANT APPLICATIONS

1. Before constructing the circuit of Figure 21-10, describe (in a short statement) how the relay will function when the voltage is applied.
2. Construct the circuit. Close the switch. Were your predictions correct?

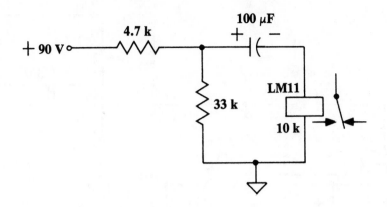

FIGURE 21-10

3. Repeat Steps 1 and 2 for Figure 21-11. What will the light do when the voltage is applied to the circuit? What effect will varying the 1 MΩ pot have on the circuit?

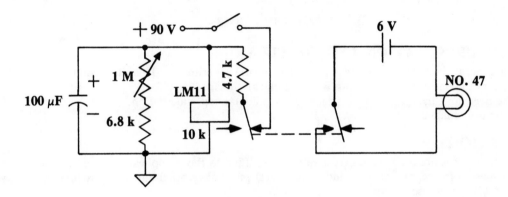

FIGURE 21-11

4. Assume the NE-2 in Figure 21-12 fires at 90 volts. Mathematically determine the time lapse between the closing of the switch and the firing of the NE-2.

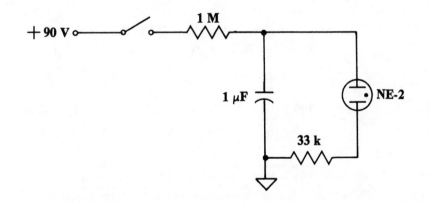

FIGURE 21-12

5. Construct the circuit of Figure 21-12, close the switch, and measure the time until the neon fires. If this step is repeated, open the switch and then discharge the capacitor.
6. Compare the two firing times.
7. Construct the circuit of Figure 21-13. Determine the cycle time of the light.

149

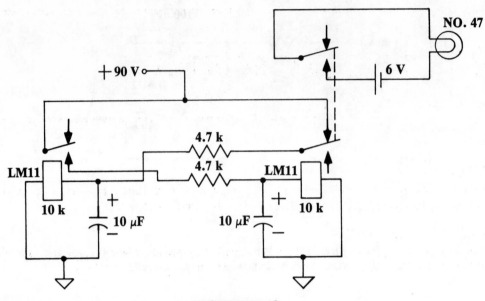

FIGURE 21-13

DATA INTERPRETATION AND CONCLUSIONS

Write a general summary of the ideas presented in this experiment. This discussion should include:
1. Factors to consider when making RC time calculations.
2. The need for "ball park" calculations prior to actual experimentation.
3. Your own conclusions.

APPLICATIONS

RC timing devices are used quite extensively. In addition to those applications used in the examples of this experiment, other applications would include: (1) pulse shaping circuits, (2) frequency selective networks, and (3) special oscillators.

PROBLEMS

1. Draw the circuit of the charge path for the network of Figure 21-14.
2. Draw the discharge path for the network of Figure 21-14.
3. Assume that a 10 kΩ relay is connected across the capacitor and repeat problems 1 and 2. R_p is equal to 70 kΩ for this problem.

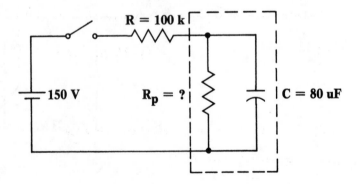

FIGURE 21-14

150

22 | BRIDGE CIRCUITRY

OBJECTIVES
The bridge circuit is used widely in electrical measurement and control circuitry. In this experiment, you are introduced to bridge circuit principles and bridge circuit applications.

THE BRIDGE NETWORK
The basic bridge network is shown in Figure 22-1. The detector voltage (E_d) is zero whenever the bridge is in balance. Equation 22-1 defines a balanced bridge in terms of ratios.

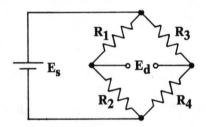

where: E_s = source voltage
E_d = detector voltage
R_1 through R_4 = resistance of arms

FIGURE 22-1

$$\frac{R_1}{R_2} = \frac{R_3}{R_4}$$

EQUATION 22-1

By making one of the resistance arms adjustable, it is possible to compensate for a change in the resistance of another arm. When the principle of compensating through rebalance is applied to the circuit of Figure 22-2, the equation can be rewritten as Equation 22-2.

$$\frac{R_1}{R_x} = \frac{R_3}{R_4}$$

EQUATION 22-2

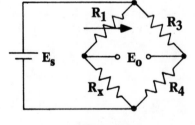

Rearranging this equation to solve for R_x is shown in Equation 22-3.

$$R_x = \frac{R_1 R_4}{R_3}$$

EQUATION 22-3

FIGURE 22-2

EXAMPLE 1:
Determine the value of R_x for a detector voltage of zero volts when: $R_1 = 6\ \Omega$, $R_3 = 8\ \Omega$, and $R_4 = 12\ \Omega$.

SOLUTION: Substitute into Equation 22-3.

$$R_x = \frac{6 \times 12}{8} = 9\ \Omega$$

ACCURACY OF THE BRIDGE CIRCUIT
Reflecting back to Experiment 1, recall that the resistance scale on the VOM is non-linear. Furthermore, it was established that the degree of accuracy was dependent upon (1) the operator's ability to properly extract the data and (2) the portion of the meter scale that was being used. Further consideration

will now be given to this problem, first in terms of direct meter reading, and then in terms of a bridge measurement.

Figure 22-3 shows a circuit designed to register incremental changes in resistance. The variable power supply is initially adjusted for a current of 1 mA with all the switches closed. The power supply is adjusted for a voltage of 1.0 volt. The new current level can now be computed as each switch is opened. These currents are listed in Table 22-1.

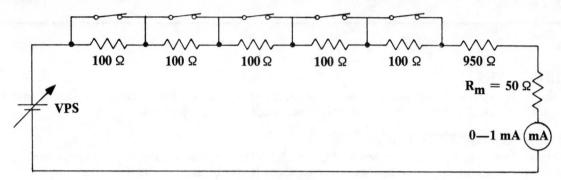

FIGURE 22-3

R_T	I_T
1.0 kΩ	1.000 mA
1.1 kΩ	0.909 mA
1.2 kΩ	0.833 mA
1.3 kΩ	0.769 mA
1.4 kΩ	0.714 mA
1.5 kΩ	0.667 mA

TABLE 22-1

When these currents are plotted on the meter scale of Figure 22-4, extreme crowding takes place and, as a result, the accuracy of the reading is impaired.

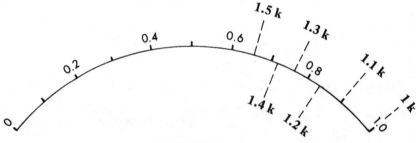

FIGURE 22-4

When the same set of resistors are used as the unknown arm of a bridge, see Figure 22-2, the variable, or adjustable, arm can rebalance the network. With this type of circuit, the accuracy of the meter is unimportant since the reading is not taken from a meter scale. The meter serves only to indicate when a balance exists. This, of course, does not completely eliminate the operator error since judgment is still required in determining when a null, or balance, is indicated by the meter. However, it is a small error.

When the unknown resistance changes at a fairly rapid rate, it becomes impractical to try to keep the bridge balanced. In such a situation, the detector voltage (E_d) can be utilized. The magnitude of E_d can be determined by Equation 22-4 if the detector resistance is at least 50 times greater than the internal resistance of the bridge. The equation is developed around the circuit of Figure 22-1.

$$E_d = E_s \left[\frac{R_2R_3 - R_1R_4}{(R_1 + R_2)(R_3 + R_4)} \right]$$

EQUATION 22-4

152

EXAMPLE 2:

Determine the detector voltage indicated by a MFM as each switch is opened in the network of Figure 22-5.

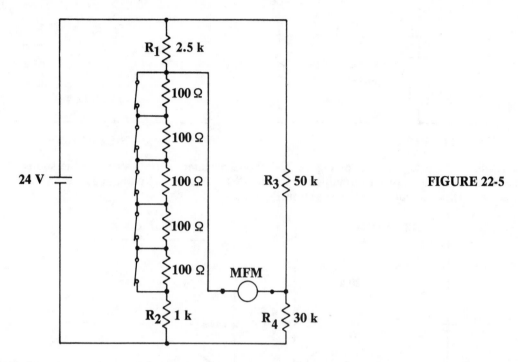

FIGURE 22-5

SOLUTION: Initially the arm representing R_2 will consist only of a 1 kΩ resistor. By Equation 22-4:

$$E_d = 24\left(\frac{1\ k\Omega \times 50\ k\Omega - 2.5\ k\Omega \times 30\ k\Omega}{(2.5\ k\Omega + 1\ k\Omega) \times (30\ k\Omega + 50\ k\Omega)}\right) = -2.14\ V$$

Table 22-2 lists all computed detector voltages for the circuit.

R_2	E_d	ΔE_d
1.0 kΩ	− 2.14 V	0.47
1.1 kΩ	− 1.67 V	0.46
1.2 kΩ	− 1.21 V	0.42
1.3 kΩ	− 0.79 V	0.41
1.4 kΩ	− 0.38 V	0.38
1.5 kΩ	0	

R_2	E_d	ΔE_d
11.0 kΩ	− 2.38 V	0.48
11.1 kΩ	− 1.90 V	0.48
11.2 kΩ	− 1.42 V	0.47
11.3 kΩ	− 0.95 V	0.48
11.4 kΩ	− 0.47 V	0.47
11.5 kΩ	0	

TABLE 22-2   **TABLE 22-3**

NOTE: Observe that the incremental change (ΔE_d) shows more linearity and less crowding.

If the resistance values for the arms of R_1 and R_2 were increased, as shown in Figure 22-6, the linearity becomes nearly perfect. See Table 22-3.

There are, of course, instances when it is more economical to use a low resistance meter as an integral part of the circuit. Some design considerations for this type of circuit are given in Example 3.

EXAMPLE 3:

A thermistor is being used to measure temperature. The thermistor resistance is 12 kΩ at 20°C and decreases at the rate of 25 Ω / ° C. The maximum temperature that is to be measured is 40°C. A 0 - 15 µA meter with an internal resistance of 4.5 kΩ will be used. Determine (1) the source voltage required and (2) the meter calibration points for 4°C changes.

153

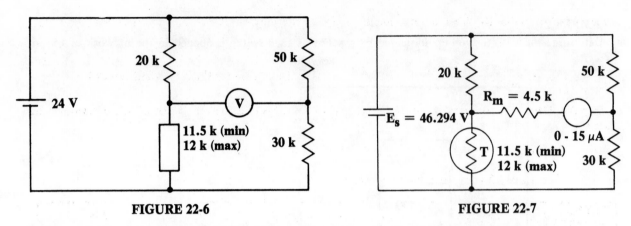

FIGURE 22-6   **FIGURE 22-7**

SOLUTION: The bridge circuit of Figure 22-7 shows no internal resistance for the source, and no resistance is shown in the branch containing the source. When a bridge circuit appears in this form, it can be reduced to a Thevenin equivalent as shown in Figure 22-8.

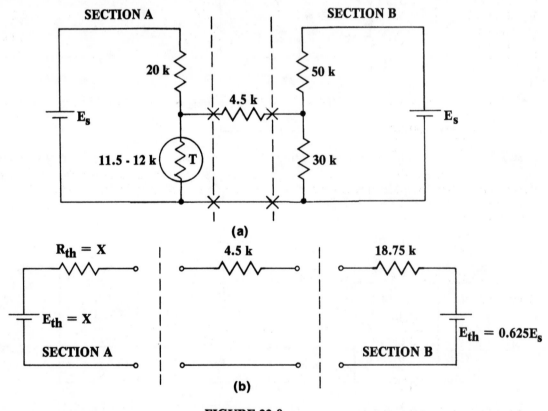

FIGURE 22-8

Step 1.   Determine R_{th} for Section B.

$$R_{th} = \frac{30 \text{ k}\Omega \times 50 \text{ k}\Omega}{30 \text{ k}\Omega + 50 \text{ k}\Omega} = 18.75 \text{ k}\Omega$$

Step 2.   Determine E_{th} for Section B.

$$E_{th} = E_s(50/80) = 0.625E_s$$

Step 3.   Determine the resistance change in the thermistor. Since the resistance change is specified as 25 $\Omega/°$ C, a 4 degree change would be equal to 100 ohms.

154

Step 4.   Determine R_{th} for Section A for the following thermistor values: 11.5 kΩ, 11.6 kΩ, 11.7 kΩ, 11.8 kΩ, 11.9 kΩ, and 12 kΩ.

$$R_{th} = \frac{11.5 \text{ k}\Omega \times 20 \text{ k}\Omega}{11.5 \text{ k}\Omega + 20 \text{ k}\Omega} = 7.30 \text{ k} \qquad \text{(See Table 22-4 for the remaining values.)}$$

Step 5.   Determine the total circuit resistance of Figure 22-8(b) using the values of R_{th} determined in Step 4.

$$R_T (11.5 \text{ k}\Omega) = 7.30 \text{ k}\Omega + 4.5 \text{ k}\Omega + 18.75 \text{ k}\Omega = 30.55 \text{ k}\Omega$$

(See Table 22-4 for the remaining values.)

THERMISTOR RESISTANCE	R_{th} (A)	R_{total}	E_{th} (A)	I_T	ΔI_T
11.5 kΩ	7.301 kΩ	30.551 kΩ	29.3904 V	15.000 μA	3.032 μA
11.6 kΩ	7.341 kΩ	30.591 kΩ	29.2974 V	11.968 μA	3.016 μA
11.7 kΩ	7.382 kΩ	30.632 kΩ	29.2055 V	8.951 μA	3.021 μA
11.8 kΩ	7.421 kΩ	30.671 kΩ	29.1132 V	5.931 μA	2.970 μA
11.9 kΩ	7.460 kΩ	30.710 kΩ	29.0222 V	2.960 μA	2.960 μA
12.0 kΩ	7.500 kΩ	30.750 kΩ	28.9313 V	0	

TABLE 22-4

Step 6.   Using Equation 22-1, it can be shown that, when the thermistor is 12 kΩ, the bridge is balanced.

$$\frac{20 \text{ k}\Omega}{12 \text{ k}\Omega} = \frac{50 \text{ k}\Omega}{30 \text{ k}\Omega}$$

Step 7.   When the thermistor equals 11.5 kΩ, the "worst case" of bridge unbalance occurs. The circuit would offer minimum resistance and, at the same time, have a maximum detector voltage. Thus, a maximum current would also flow. Since the maximum current that can flow is 15 μA, the source voltage can be computed as follows:

$$E_{th}(A) - E_{th}(B) = E(\text{differential})$$

$$E(\text{diff}) = 30.55 \text{ k}\Omega \times 15 \text{ }\mu\text{A} = 0.458265 \text{ V}$$

Therefore, since

$$E_{th}(A) = \frac{E_s 20}{31.5} \qquad \text{and} \qquad E_{th}(B) = \frac{E_s 50}{80}$$

$$E(\text{diff}) = \frac{E_s 20}{31.5} - \frac{E_s 50}{80} = 0.458265 \text{ V}$$

Factoring our E_s and solving:

$$E_s = \frac{0.458265}{0.6349 - 0.625} = 46.29 \text{ V}$$

Step 8.   Determine $E_{th}(B)$.

$$E_{th} = 46.29 \times 0.625 = 28.93 \text{ V}$$

Step 9.   Determine $E_{th}(A)$ for each thermistor change.

$$E_{th} = \frac{46.29 \times 20}{31.5} = 29.39 \text{ V} \qquad \text{(See Table 22-4 for the remaining values.)}$$

155

Step 10. Determine the circuit current for each thermistor change.

$$E_{th}(A) - E_{th}(B) = E(diff)$$

$$I_T = \frac{E(diff)}{R_T} = \frac{29.39 - 28.93}{30.55 \text{ k}\Omega} = \frac{0.460}{30.55 \text{ k}\Omega} = 15 \text{ }\mu A$$

Step 11. Calibrate the meter scale. ΔI_T of Table 22-4 gives an indication of the degree of linearity.

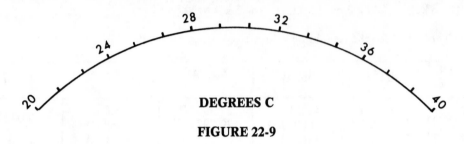

DEGREES C

FIGURE 22-9

SUGGESTED OBSERVATIONS
1. Protect the thermistor from air currents.
2. Record all measurements, observations, and calculations in the data tables.

LIST OF MATERIALS

Alternate Materials

1. MFM
2. VOM
3. VPS
4. Thermistor 41D2
5. SPST Switch
6. 1 MΩ Potentiometer
7. Resistors—all at least ½ watt

| 2.7 kΩ | 10 kΩ(2) | 47 kΩ |
| 4.7 kΩ | 15 kΩ | 470 kΩ |

_____

_____

_____

_____

EXPERIMENTAL PROCEDURE
In this experiment, you will construct and evaluate bridge circuitry and detector linearity.

SECTION A: BRIDGE CIRCUIT PRINCIPLES
1. Construct the circuit of Figure 22-10. Use a MFM as the balance detector. Before placing the analog voltmeter in the circuit, move the meter pointer from the left zero position to the "center zero" (center of the meter scale). Use the zero adjust control to accomplish this. The meter can now provide both negative and positive indications. (No adjustment required for DVM.)

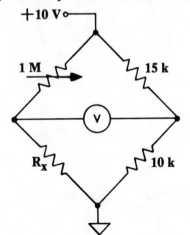

FIGURE 22-10

2. Place an unknown resistance (from 4.7 kΩ to 470 kΩ) in the bridge as R_x. Adjust the potentiometer so the pointer is again at "center zero".

3. With the power off, carefully remove the potentiometer from the circuit and measure its resistance. Use Equation 22-3 and the component values of Figure 22-10 to determine the value of R_x.

4. With an ohmmeter, measure the value of R_x.

6. Repeat Steps 2 through 5 for an additional unknown resistance.

SECTION B: BRIDGE CIRCUIT CALCULATION

1. Construct the circuit of Figure 22-11. Use a VPS as the source. With the analog voltmeter center zeroed, balance the bridge. Remove the 1 MΩ potentiometer and measure its value for the balanced bridge.

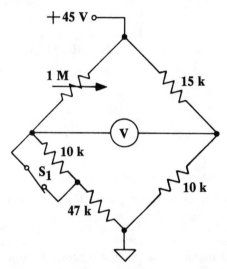

FIGURE 22-11

2. Using the measured value of the potentiometer determined in Step 1 and the remaining circuit values, mathematically determine (using Equation 22-4) the value and direction of the voltmeter deflection when S_1 is opened.

3. With the bridge balanced, open S_1 and record the direction and magnitude of the voltage. (Remember that the center of the analog meter scale is now the zero point and all readings are referenced to this point.)

4. Compare the measured voltage to the calculated voltage (Steps 2 and 3).

SECTION C: BRIDGE CIRCUIT CALCULATIONS

1. Thevenize the circuit of Figure 22-12 and determine the amount and the direction of the current passing through the 2.7 kΩ resistor.

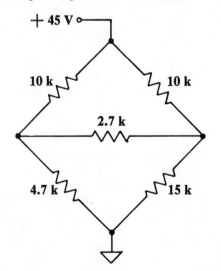

FIGURE 22-12

157

2. Construct the circuit and measure the current through the 2.7 kΩ resistor.
3. Compare the calculated and measured values of current (Steps 1 and 2).

SECTION D: BRIDGE CIRCUIT APPLICATION

1. Construct the circuit of Figure 22-13. With the analog voltmeter center-zeroed, balance the bridge.

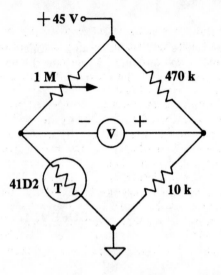

FIGURE 22-13

2. Warm the thermistor by placing it between your thumb and forefinger. In a short statement describe the balance detector's movement.
3. Heat the thermistor by placing a 30 to 60 watt soldering iron ¼ to ½ inch beneath the thermistor. Don't directly heat the thermistor. Describe the balance detector's movement.
4. If a component cooler ("Zero-mist", "Freez-mist") is available, cool the thermistor. Handle this material carefully. Describe the balance detector's movement.

DATA INTERPRETATION AND CONCLUSIONS
Write a general summary of the ideas presented in this experiment. This discussion should include:
1. Your own conclusions.

APPLICATIONS
See Problem No. 1.

PROBLEM
1. List several applications of the bridge circuit.

A-1   SOLDERING TECHNIQUES

Soldering is only one of several methods used to join wires and components to terminals. This process, however, is used extensively in the electronics industry. The soldering process can be accomplished by several different techniques, one of which is hand soldering. Many of the "production line" items formerly requiring hand soldering are now being "machine" soldered. Regardless of the great strides made in production soldering, rework and repair work have to be done by hand.

To understand why certain soldering techniques were developed, a knowledge of the physical properties of solder is necessary. Solder (an alloy) is a mixture of two metals, tin and lead. One of the useful properties of an alloy is that it melts at a temperature lower than that of the pure metals. Figure A 1-1 shows that pure lead melts at 621°F and tin at 450°F. Note, however, that when these two metals are combined, the alloy melts at a temperature lower than that of either tin or lead. The lowest melting point for any alloy is called the **eutectic** point. (Pronounced u-tek-tik.) For a mixture of 63% tin and 37% lead, the eutectic point is 361° F. It can then be stated that the eutectic temperature is 361°F and that the eutectic composition is that it lacks a plastic state. This means that as the solder melts it goes from a solid to a liquid state instantly. As the alloying mixture deviates from the eutectic composition, say 50% tin and 50% lead, a plastic state appears. In other words, as the 50/50 solder is heated, it goes from hard to soft and soft to liquid. The time that it remains in the soft state depends upon the alloying mixture. The reverse process is followed as the solder begins to cool.

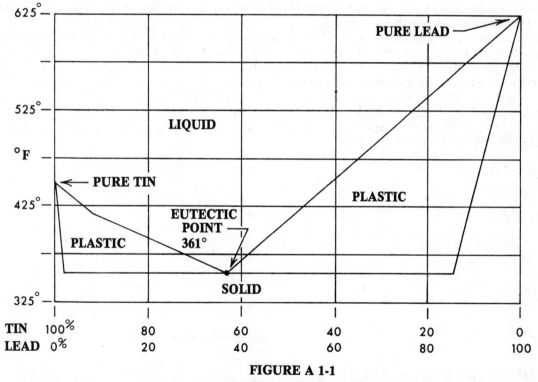

FIGURE A 1-1

Since heat in general effects the performance of electronic components, a low soldering temperature is desirable. A commonly used alloy is 60/40 (60% tin and 40% lead) with a melting point of 370°F. Since this is only an increase of 9°F above the eutectic point of 361°F, there is a relatively small plastic range. Keep in mind that as the solder is cooling, movement of the connection while in the liquid state is not harmful, but movement while in the plastic state results in a "cold" solder joint. The surface of the cold solder joint usually has a very dull, rough texture and must be avoided. If the connection is moved while in this state, **reheat** the joint and allow it to cool undisturbed!

During the soldering process a chemical action takes place between the solder and the metals being joined. In order to insure that this action takes place, the metals being joined must be cleaned of all foreign matter, dirt, grease, etc. Even though the connection has been cleansed of foreign matter, the surface of the metal develops an oxide coating. As long as this oxidation remains, the material cannot be soldered. An agent for removing oxidation is called **flux**. Solder flux is available in either liquid or paste form and is classified as being acid or rosin types. Rosin flux is used exclusively for electronic circuit wiring; ACID flux is **never** used with electronic wiring.

In order to insure that proper bonding takes place, it is necessary to raise all the metals being soldered to the melting temperature of the solder. It is imperative, then, that a soldering iron of sufficient **wattage** be used. Several wattage ratings are available for both soldering irons and soldering guns. The lightweight, 20 to 60 watt, soldering pencil is usually sufficient for soldering wires and components to lugs, sockets, and printed wiring boards. The soldering gun is used principally in repair work where it is impractical to keep a soldering iron heated.

Copper, being one of the better conductors of heat and being inexpensive, is used extensively for soldering iron tips. The use of copper alone as a tip usually means considerable maintenance of the tip, such as filing to remove pits, etc. To overcome this weakness of copper and yet retain heat conductivity, a thin layer of iron is placed over the tip. Iron tips, unlike copper tips, will show little deterioration even after months of continuous use.

Now that the technical aspects of soldering have been presented, a partial list of the most common causes of poor solder joints is given.

1. Lack of heat—cold solder joint.
2. Dirty tip—solder won't flow.
3. Dirty connections— failure to remove foreign material.
4. Connection moved prior to solidification of the solder.
5. Rosin joint—connection appears to have been soldered, but a layer of rosin is between the wires preventing their contact—a high electrical resistance joint.

Since heat can be quite destructive to some electrical components, a method of removing some of the heat from the components must be provided. One of the simplest ways of shunting heat from the component is with the use of an inexpensive "heat sink". The heat sink is similar to a pair of tweezers, except that in this case compressing the tongs will release the component. To use the heat sink, merely clip it on the lead of the component.

Within the limits of this experiment, an opportunity is given to learn how to make good solder connections. In order to insure that the proper techniques are developed, your instructor will be assisting you.

SUGGESTED OBSERVATIONS

1. Keep the soldering iron wiped clean.
2. Heat the connection before attempting to solder.
3. Do not shake the soldering iron to remove excess solder; wipe it clean.
4. Use care when soldering. People, clothing, components, and even tables can be burned or marred because of carelessness.
5. **NEVER** file an iron-plated soldering tip. File only copper tips.

LIST OF MATERIALS

Alternate Materials

1. Soldering Iron
2. Rosin Core Solder
3. Long Nose Pliers _____
4. Diagonal Cutters
5. Wire Strippers _____
6. 8 Feet Hook-up Wire
7. Two Mounted Tube Sockets

Multipinned Connectors _____

EXPERIMENTAL PROCEDURE

SECTION A: PREPARATION FOR SOLDERING

The following procedure is used for iron clad tips.

1. While soldering iron is heating, test the tip temperature with a piece of rosin core solder. The iron is ready to tin when the solder melts.
2. Once the iron is hot, tin the tip by coating the end with solder. The flux in the solder will remove any oxides.
3. Wipe the tip with a cloth or damp sponge to remove the excess solder. The iron should now appear bright and clean. The iron is now ready for use.
4. Keep the tip clean by wiping it with a rag or a commercially made sponge tip cleaner.

SECTION B: SOLDERING

Read the procedure (Steps 1 through 5) before attempting to solder. The time needed to make a solder connection is about ten seconds.

1. Strip about ¾ inch of the insulation from eight 3-inch pieces of hook-up wire.
2. Connect one wire to a terminal of the tube socket. This wire is usually secured to the lug by making a small hook in the end and then crimping the wire around the lug with long nose pliers. Since the wire will need to be removed, do not make a hook or crimp the wire—only place it through the terminal.
3. The clean, hot, soldering iron is placed in contact with the wire and the terminal; a small amount of solder is added to the area where the iron is in contact with the work. This solder provides a heat bridge for the transfer of heat from the tip to the connection.
4. Once the joint reaches soldering temperature, additional solder is applied to the work: **NEVER** to the iron. Only enough solder is added to fill the joint and cover the wire.
5. Allow the solder to flow around the wire and the flux to volitalize. The iron is removed and the joint is left undisturbed to cool. The flux residue is nonconductive and may be left. However, it can be removed with a commercial flux remover.
6. Repeat Steps 2 through 5 for each of the remaining terminals of the tube socket.
7. When the tube socket is completely soldered, have the instructor inspect it.

SECTION C: UNSOLDERING

Once the instructor approves the soldering, unsolder the connections and clean up the tube socket.

1. Heat the joint, remove the wire, and allow the solder to flow onto the iron. Wipe the solder off the iron. **DO NOT** shake it off as it might burn someone. **Solder wick** may be used to aid in removing the solder from the terminal.
2. Repeat this process until the socket is clean. Have the instructor inspect the socket.

SECTION D: WIRING

1. Connect a wire from one tube socket terminal to a terminal on a second socket. Remember, do not hook or crimp the wire.
2. Repeat this for each tube socket terminal. Have the instructor inspect your work.
3. Once your work is satisfactory, unsolder and clean the tube sockets. Have the instructor inspect your work.

SECTION E: SHOP JOB

Ask the instructor if he has anything that needs to be built up or stripped down.

DATA INTERPRETATION AND CONCLUSIONS

Write a general summary of the ideas presented in this experiment. This discussion should include:
1. The reason for using certain alloys of solder for electronic soldering.
2. The reasons for using rosin flux.
3. The need for adequate heat while soldering.
4. A brief review of the steps needed to make a solder connection.
5. Your own conclusions.

APPLICATIONS

Soldering is used in the construction of: (1) radios, (2) televisions, (3) motors, (4) generators, (5) tape recorders, and (6) amplifiers.

PROBLEMS

1. Using an industrial parts catalog, list the manufacturer, model, and cost of three soldering irons. Include the soldering pencil as one of the three irons. (Consult Appendix A-2 for industrial parts catalog usage.)
2. Determine the cost of the following items:
   a. Heat sinks for soldering.
   b. Long nose pliers—small and large.
   c. Diagonal cutters—small and large.
   d. One pound each of 40/60, 50/50, and 60/40 solder.
3. If given a choice between 60/40 solder and 50/50 solder, which would be chosen and why?

A-2 INTERPRETING THE INDUSTRIAL PARTS CATALOG

Engineering technicians are faced with the perpetual task of determining the availability, cost, physical size, tolerance, and other specifications pertaining to electronic components, hardware, and devices. This data is available from manufacturers in the form of individual "spec" (specification) sheets or through industrial parts catalogs. The **electronics part catalog** is published for industrial use by several of the large electronic parts distributors. The industrial parts catalog is of such tremendous assistance that you will find nearly every experiment in this manual referring to it.

Since several manufacturers market the same product (e.g. resistors), they like to include pictures of their product. These pictures are used extensively throughout the parts catalog, and they are a great aid to the buyer.

Another feature of this catalog is the way in which it is structured. Note that all the different components are grouped. Thus, if the index were used to determine where information on resistors could be found, a certain span of page numbers would be listed.

If the catalog does not contain all of the information required, one can write directly to the manufacturer and request a specification sheet. Specification sheets are also made available through the "reader service" section of trade magazines and journals.

The names and addresses of several parts suppliers that have catalogs are listed in this exercise. These catalogs are supplied at little or no cost upon written request. In addition, the addresses of some trade magazines and journals, which are of benefit to both the engineer and engineering technician, are also provided.

SUGGESTED OBSERVATIONS

1. Observe the structural layout of at least two parts catalogs.
2. While doing this experiment, try to associate the physical components with the pictures in the catalog.
3. In an effort to make work as neat as possible, observations should first be entered in a rough data table (of your construction) and then the finished work should be transferred to the data tables of this experiment.

LIST OF MATERIALS

Alternate Materials

1. Industrial Parts Catalog
2. Battery
3. Capacitor
4. Diode
5. Inductor
6. Lamp
7. Meter
8. Potentiometer
9. Relay
10. Resistor
11. Switch

_____

_____

_____

EXPERIMENTAL PROCEDURE

SECTION A: CATALOG FAMILIARIZATION

In this section the concern will be with identifying the catalog structure. Place all answers in the data table.

1. Determine the name of the components issued by identifying them with pictures and specifications in the parts catalog. Record the name in the data table.
2. Have the instructor check the names you have listed.

SECTION B: COMPONENT PRICING

1. In the data table, record the name and catalog number of the industrial parts catalog being used.
2. Determine and record the price of each of the components. (By manufacturer, if possible.)
3. In addition to the cost of the component, record the stock number, manufacturer number, and description (including dimensions, values, tolerance, etc.).
4. For each of the recorded components, list an additional manufacturer who supplies a similar

component. If possible, list manufacturers that have components with equivalent ratings. Repeat Steps 2 and 3.

SECTION C: IDENTIFICATION
In this section you will be tested by your instructor to see if the components can be identified by sight.
1. When you are confident of the components' identity, take them to the instructor so he may test your knowledge of them. Be sure he initials the data table.

DATA INTERPRETATION AND CONCLUSIONS
Write a short summary of how the parts catalog is used. This summary should include:
1. A description of how components are grouped in the catalog.
2. A comment on the informational value of figures and component descriptions.
3. A discussion of any difficulties encountered while using the catalog.
4. Your own conclusions.

APPLICATIONS
The parts catalog is consulted when: (1) estimating cost, (2) determining component availability, (3) selecting standard values, and (4) determining dimensions.

PROBLEMS
1. Is there any advantage to buying in quantities? If so, what is it?
2. What is the purpose of a cabinet or a box chassis?
3. What do you think a "Relay Rack" could be used for?

PARTS CATALOGS

ALLIED ELECTRONICS CORPORATION
100 W. Western Avenue
Chicago, Illinois 60680

NEWARK ELECTRONICS CORPORATION
223 W. Madison Street
Chicago, Illinois 60606

BURSTEIN-APPLEBEE COMPANY
3199 Mercier Street
Kansas City, Missouri 64111

OLSON ELECTRONICS, INC.
440 S. Forge Street
Akron, Ohio 44308

LAFAYETTE RADIO ELECTRONICS
111 Jericho Turnpike
Syosset, Long Island, New York 11791

RADIO SHACK
2727 West 7th Street
Fort Worth, Texas 76107

TRADE MAGAZINES AND JOURNALS

CQ RADIO AMATEUR'S JOURNAL
Cowan Publishing Corp.
67 West 44th Street
New York, New York 10036

POPULAR ELECTRONICS
Ziff-Davis Publishing Co.
Portland Place
Boulder, Colorado 80302

ELECTRICAL DESIGN NEWS
Electrical Design News
3375 S. Bannock
Englewood, Colorado 80110

QST (AMATEUR RADIO)
American Radio Relay League
225 Main Street
Newington, Connecticut 06111

ELECTRONIC TECHNICIAN
Electro Tech
1 East First Street
Duluth, Minnesota 55802

RADIO—ELECTRONICS
Gernsback, Inc.
Ferry Street
Concord, New Hampshire 03302

73 MAGAZINE (AMATEUR RADIO)
73, Inc.
Peterborough, New Hampshire 03458

A-2 INTERPRETING THE INDUSTRIAL PARTS CATALOG DATA TABLES

Name _____

Date _____

Class _____

SECTION A: CATALOG FAMILIARIZATION

1. Instructor's OK _____

SECTION A Component	SECTION B Catalog =				
NAME	STOCK #	MFG.	MFG. #	DESCRIPTION	COST

SECTION C: IDENTIFICATION

1. Instructor's OK _____

B-1 EQUATIONS USED IN THE MANUAL

EXPERIMENT 1:

$$\frac{FSD \times MSA}{AD} \times 100 = \text{percentage of accuracy at the deflection point} \qquad \textbf{EQUATION 1-1}$$

$$\text{plus or minus range of deviations} = (FSC)(\pm MSA) \qquad \textbf{EQUATION 1-2}$$

EXPERIMENT 2:

$$\% \text{ difference} = \frac{MV - CCV}{CCV} \times 100 \qquad \textbf{EQUATION 2-1} \qquad\qquad R = \rho\,\frac{L}{A} \qquad \textbf{EQUATION 2-2}$$

$$\frac{R_1}{R_2} = \frac{\left(\dfrac{1}{a_t} - t\right) + T_1}{\left(\dfrac{1}{a_t} - t\right) + T_2} \qquad \textbf{EQUATION 2-3} \qquad\qquad L = \frac{RA}{\rho \times 5280 \times 2} \qquad \textbf{EQUATION 2-4}$$

EXPERIMENT 4:

$$E_T = E_1 + E_2 + E_3 + \dots + E_n \qquad \textbf{EQUATION 4-1} \qquad\qquad \text{(a)} \quad I = E/R \qquad \textbf{EQUATION 4-4}$$

$$I_T = I_1 = I_2 = I_3 = \dots = I_n \qquad \textbf{EQUATION 4-2} \qquad\qquad \text{(b)} \quad E = IR$$

$$R_T = R_1 + R_2 + R_3 + \dots + R_n \qquad \textbf{EQUATION 4-3} \qquad\qquad \text{(c)} \quad R = E/I$$

$$E_x = \frac{E_A R_x}{\Sigma R} \qquad \textbf{EQUATION 4-5} \qquad\qquad R_{int} = \left[\frac{E_s}{E_o} - 1\right] R_L \qquad \textbf{EQUATION 4-6}$$

$$\% \text{ error} = \frac{MV - CV}{CV} \times 100 \qquad \textbf{EQUATION 4-7}$$

EXPERIMENT 5:

$$\textbf{EQUATION 5-4(a)}$$

$$E_T = E_1 = E_2 = E_3 = \dots = E_n \qquad \textbf{EQUATION 5-1} \qquad G_T = \frac{1}{R_T} \qquad G_1 \frac{1}{R_1} \qquad G_2 = \frac{1}{R_2}$$

$$I_T = I_1 + I_2 + I_3 + \dots + I_n \qquad \textbf{EQUATION 5-2} \qquad\qquad\qquad\qquad\qquad \textbf{EQUATION 5-4(b)}$$

$$G_T = G_1 + G_2 + G_3 + \dots + G_n \qquad \textbf{EQUATION 5-3} \qquad R_T = \frac{1}{G_T} \qquad R_1 = \frac{1}{G_1} \qquad R_2 = \frac{1}{G_2}$$

$$E_T = \frac{I_T}{G_T} \qquad E_1 = \frac{I_1}{G_1} \qquad \text{etc.} \qquad \textbf{EQUATION 5-5(a)} \qquad R_T = \frac{R_1 R_2}{R_1 + R_2} \qquad \textbf{EQUATION 5-6}$$

$$\textbf{EQUATION 5-7}$$

$$I_T = E_T G_T \qquad I_1 = E_1 G_1 \qquad \text{etc.} \qquad \textbf{EQUATION 5-5(b)} \qquad R_T = \frac{1}{\dfrac{1}{R_1} + \dfrac{1}{R_2} + \dfrac{1}{R_3}}$$

$$I_x = \frac{I_T R}{\Sigma R} \qquad \textbf{EQUATION 5-8(a)} \qquad\qquad I_x = \frac{I_T G}{\Sigma G} \qquad \textbf{EQUATION 5-8(b)}$$

EXPERIMENT 7:

$$k_{shunt} = \frac{I_m \times R_m}{I_{shunt}} \qquad \textbf{EQUATION 7-1} \qquad\qquad R_{mult} = \frac{E}{I_m} - R_m \qquad \textbf{EQUATION 7-2}$$

EXPERIMENT 7:

$$\Omega/V = \frac{1}{I_m}\frac{V}{}\qquad \textbf{EQUATION 7-3}$$

EXPERIMENT 8:

$$R_{meq} = \frac{E_2 R_s}{E_1 - E_2}\qquad \textbf{EQUATION 8-1}$$

EXPERIMENT 9:

EQUATION 9-1

$$\text{Wattage Gradient} = \frac{MPR}{L}$$

EQUATION 9-2

(a) $P = IE$　　(b) $P = \dfrac{E^2}{R}$　　(c) $P = I^2 R$

EXPERIMENT 11:

EQUATION 11-1

$$R_{th} = \left[\frac{E_{th}}{E_o} - 1\right] R_L$$

EXPERIMENT 13:

EQUATION 13-1

$$I_L = \frac{I_{sc} R_{th}}{R_{th} + R_L}$$

EXPERIMENT 15:

EQUATION 15-1

$$E = \frac{\dfrac{E_1}{R_1} + \dfrac{E_2}{R_2} + \dfrac{E_3}{R_3} + ... + \dfrac{E_n}{R_n}}{\dfrac{1}{R_1} + \dfrac{1}{R_2} + \dfrac{1}{R_3} + ... + \dfrac{1}{R_n}}$$

EXPERIMENT 16:

EQUATION 16-1

$$K_1 = R_{11} I_1$$

EQUATION 16-2

(a) $K_1 = R_{11} I_1 + R_{12} I_2$

(b) $K_2 = R_{21} I_1 + R_{22} I_2$

EQUATION 16-3

(a) $K_1 = R_{11} I_1 + R_{12} I_2 + R_{13} I_3$

(b) $K_2 = R_{21} I_1 + R_{22} I_2 + R_{23} I_3$

(c) $K_3 = R_{31} I_1 + R_{32} I_2 + R_{33} I_3$

EQUATION 16-4

$$K_1 = G_{11} V_1$$

EQUATION 16-5

(a) $K_1 = G_{11} V_1 + G_{12} V_2$

(b) $K_2 = G_{12} V_1 + G_{22} V_2$

EQUATION 16-6

(a) $K_1 = G_{11} V_1 + G_{12} V_2 + G_{13} V_3$

(b) $K_2 = G_{21} V_1 + G_{22} V_2 + G_{23} V_3$

(c) $K_3 = G_{31} V_1 + G_{32} V_2 + G_{33} V_3$

EXPERIMENT 17:

EQUATION 17-1

$$L_T = L_1 + L_2 + ... + L_n$$

EQUATION 17-2

$$L_T = \frac{1}{\dfrac{1}{L_1} + \dfrac{1}{L_2} + \dfrac{1}{L_3} + \cdots + \dfrac{1}{L_n}}$$

EXPERIMENT 18:

EQUATION 18-1

$$C = \frac{Q}{E}$$

EQUATION 18-2

$$Q_T = Q_1 = Q_2 = Q_3 = ... = Q_n$$

EQUATION 18-3

$$C_T = \frac{1}{\dfrac{1}{C_1} + \dfrac{1}{C_2} + ... + \dfrac{1}{C_n}}$$

EQUATION 18-4

$$C_T = C_1 + C_2 + ... + C_n$$

EQUATION 18-5

$$Q_T = Q_1 + Q_2 + ... + Q_n$$

EQUATION 18-6

$$1(mA) = KC(\mu F) + \ 0.3$$

EQUATION 18-7

$$I(\mu A) = 0.003 C(\mu F) V$$

EXPERIMENT 21:

EQUATION 21-1 $\qquad$ **EQUATION 21-2** $\qquad$ **EQUATION 21-3** $\qquad$ **EQUATION 21-4**

$$E_s = E_R + E_C \qquad i = \frac{E_s}{R}e^{-t/RC} \qquad E_R = E_s e^{-t/RC} \qquad E_C = E_s(1 - e^{-t/RC})$$

EQUATION 21-5 $\qquad$ **EQUATION 21-6** $\qquad$ **EQUATION 21-7**

$$i = \frac{E_C}{R}e^{-t/RC} \qquad E_C = E_R \qquad E_R = E_C e^{-t/RC}$$

EXPERIMENT 22:

EQUATION 22-1 $\qquad$ **EQUATION 22-2** $\qquad$ **EQUATION 22-3** $\qquad$ **EQUATION 22-4**

$$\frac{R_1}{R_2} = \frac{R_3}{R_4} \qquad \frac{R_1}{R_x} = \frac{R_3}{R_4} \qquad R = \frac{R_1 R_4}{R_3} \qquad E_d = E_s\left[\frac{R_2 R_3 - R_1 R_4}{(R_1 + R_2)(R_3 + R_4)}\right]$$

C-1   GRAPHING TECHNIQUES

The graph is a very effective way of conveying information. It is constructed from values in a data table. A properly constructed and labeled graph provides a pictorial representation (in the form of a curve) of circuit conditions, values, and trends.

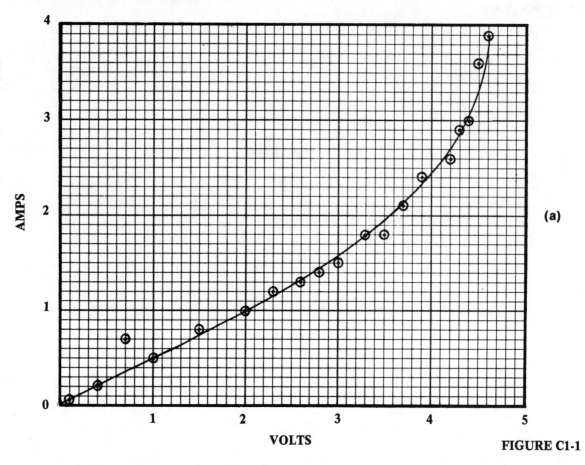

(a)

FIGURE C1-1

VOLTAGE	CURRENT	VOLTAGE	CURRENT
0.0	0.0	3.0	1.5
0.1	0.05	3.3	1.8
0.4	0.2	3.5	1.8
0.7	0.7	3.7	2.1
1.0	0.4	3.9	2.4
1.5	0.8	4.2	2.6
2.0	1.0	4.3	2.9
2.3	1.2	4.4	3.0
2.6	1.3	4.5	3.6
2.6	1.4	4.6	3.9

(b)

168

As an example, suppose that an experiment was being conducted on a circuit to determine the current requirements at different voltages. Figure C 1-1 shows the current requirements for the different voltages. Notice that the voltage, the **INDEPENDENT** quantity, is in the first column, while the current, the **DEPENDENT** quantity, is in the second column. When the data is placed in this form, there is automatically an "ordered number pair", i.e. (x, y) to locate the exact point on the graph.

To construct a graph of this data, use the following steps:

1. Make up a table of values with the independent variable placed in the first column and the dependent variable placed in the second column. Now determine the range of each variable. In Figure C 1-1, the range is from 0 to 4.6 volts and 0 to 3.9 amperes; therefore, a range of 0 to 5 and 0 to 4 was used for these values.

2. Whenever practical, use either 10 × 10 or 20 × 20 squares-to-the-inch graph paper, as this will enable the decimal part of a unit to be plotted. For this example, 10 × 10 squares-to-the-inch graph paper was used.

3. Leave sufficient room for a margin. The margin in the example contains both the range and units of the axis.

4. The assignment of unit value to the divisions of the graph requires considerable judgment. The graph of Figure C 1-1 is to be considered a minimum size. Do not plot the curve in too small an area as accuracy and reliability will be lost and will render the curve useless.

5. The curve is plotted using the "ordered number pairs" from the data table. **EACH** point has a circle around it. If two or more curves are used in the same graph, use a different symbol, such as squares, diamonds, or triangles, to identify the curve to which the point belongs. In addition, a **key** to the graph's purpose must be provided.

    The first ordered number pair is (0, 0), the second is (0.1, 0.05), the third is (0.4, 0.2), etc. In the actual plot this means that the first point is located 0 units to the right of the origin and 0 units up from the origin. The second point is located 0.1 units to the right and 0.05 units up. The third point is located 0.4 units to the right and 0.2 units up. The eleventh point is 3 units to the right and 1.5 units up, etc.

6. The final step is to draw a **smooth** curve through the plotted points. Since many of the points may not fall in line with the smoothly drawn curve, there may be a desire to erase them. **DON'T!** These points still represent laboratory measurements and as such, they must remain.

7. As a final note, when the curve is linear several points are plotted to maintain the accuracy of the function being plotted. However, when the curve is curvalinear, then many points are needed to insure the accuracy of the plotted function.

To construct a "compressed" graph of this same data, it would be necessary to use log-log or semilogarithmic graph paper. Figure C1-2 shows 2 cycle, semi-log graph paper consisting of a logarithmic axis and 10 by 10 to the inch linear axis. Special rules must be used when working with the logarithmic axis.

1. The axis cannot be started with a zero.

2. A number, such as 0.01, 0.1, 1, 10, etc., **must** start the **FIRST** vertical line when reading from **LEFT** to **RIGHT**.

3. Which number is assigned to this line, coupled with the largest number to be used, will determine the number of cycles required.

4. The logarithmic axis can be used as either the **DEPENDENT** or **INDEPENDENT** variable.

NOTE: The starting point value is purely a judgment factor on your part.

EXAMPLE 1:

If the starting point is chosen as 0.01, what is the starting value of the second cycle?

SOLUTION: $0.01 \times 10 = 0.1$

EXAMPLE 2:

If 0.001 is chosen as the starting point, what is the starting value of the fifth cycle?

SOLUTION:

First cycle starting value = 0.001
Second starting value = $0.001 \times 10 = 0.01$
Third starting value = $0.01 \times 10 = 0.1$

Fourth cycle starting value = $0.1 \times 10 = 1.0$
Fifth starting value = $1.0 \times 10 = 10$
END of fifth cycle = $10 \times 10 = 100$

Examine Figure C1-2 closely. The authors chose 0.1 as the starting value because it was judged to be adequately close to zero. Also, the largest value expected is 4.6. From these values, it was determined that 2 cycles would be adequate.

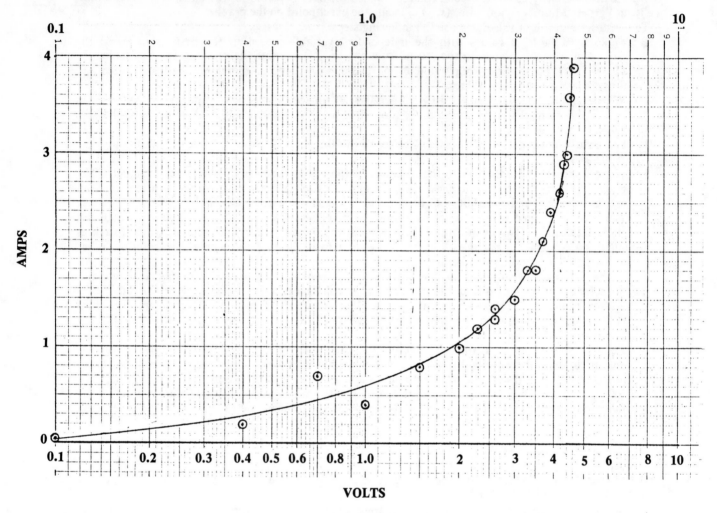

FIGURE C1-2

170

C-2 DETERMINANTS

The solution of simultaneous equations can be accomplished in many ways, one of which is through the use of determinants. The ease with which determinants can be used with loop and nodal equations is demonstrated within the structure of this manual as well as in many textbooks. Therefore, a short review is in order. If a more formal presentation is desired, consult an electronics math book or a college algebra book.

Since the primary concern is with loop and nodal equations, a standard form for each is provided.

FOR LOOP EQUATIONS:

$$K_1 = R_{11}I_1 + R_{12}I_2 + R_{13}I_3$$

EQUATION 16-3

$$K_2 = R_{21}I_1 + R_{22}I_2 + R_{23}I_3$$

$$K_3 = R_{31}I_1 + R_{32}I_2 + R_{33}I_3$$

where: K_1, K_2, K_3 = loop voltage

R_{11}, R_{22}, R_{33} = mesh resistance

$R_{12}, R_{13}, R_{21}, R_{23}, R_{31}, R_{32}$ = mutual resistance

I_1, I_2, I_3 = loop currents

FOR NODE EQUATIONS:

EQUATION 16-6

$$K_1 = G_{11}V_1 + G_{12}V_2 + G_{13}V_3$$

$$K_2 = G_{21}V_1 + G_{22}V_2 + G_{23}V_3$$

$$K_3 = G_{31}V_1 + G_{32}V_2 + G_{33}V_3$$

where: K_1, K_2, K_3 = node currents

G_{11}, G_{22}, G_{33} = mesh conductances

$G_{12}, G_{13}, G_{21}, G_{23}, G_{31}, G_{32}$ = mutual conductances

V_1, V_2, V_3 = node voltages

EXAMPLE 1:

Suppose there is a two-loop network which requires the solution of the following simultaneous equations in order to determine the value of I_1 and I_2.

$$16 = 5I_1 + 2I_2$$

$$3 = 3I_1 - 1I_2$$

Solve for I_1 and I_2 using determinants.

SOLUTION: Since there are two equations and two unknowns, only the following portion of Equation 16-3 is used.

$$K_1 = R_{11}I_1 + R_{12}I_2$$

$$K_2 = R_{21}I_1 + R_{22}I_2$$

Using these equations as the standard form to be followed, the original equations must be checked to determine if they are in standard from.

Since the K values (16 and 3) are on the left hand side of the equality sign, the I coefficients to the right, and the I_1's and I_2's in the column, the equations are in standard form.

Since the original circuit is not available to ascertain which values apply to loop 1, etc., do not be concerned with which value (16 or 3) is K_1. Hence, if $K_1 = 16$, then $R_{11} = 5$ and $R_{12} = 2$.

The standard determinant array for two equations and two unknowns is:

$$I_1 = \frac{\begin{vmatrix} K_1 & R_{12} \\ K_2 & R_{22} \end{vmatrix}}{\begin{vmatrix} R_{11} & R_{12} \\ R_{21} & R_{22} \end{vmatrix}} \qquad I_2 = \frac{\begin{vmatrix} R_{11} & K_1 \\ R_{21} & K_2 \end{vmatrix}}{\begin{vmatrix} R_{11} & R_{12} \\ R_{21} & R_{22} \end{vmatrix}}$$

When the determinant array contains only two equations, it is called a second-order determinant. The solution of a second-order determinant is obtained through a "cross multiplying" technique. Observe the following steps:

Step 1. Place the equations in standard form and substitute into the determinant array. Thus:

$$I_1 = \frac{\begin{vmatrix} 16 & 2 \\ 3 & -1 \end{vmatrix}}{\begin{vmatrix} 5 & 2 \\ 3 & -1 \end{vmatrix}} \qquad I_2 = \frac{\begin{vmatrix} 5 & 16 \\ 3 & 3 \end{vmatrix}}{\begin{vmatrix} 5 & 2 \\ 3 & -1 \end{vmatrix}}$$

Step 2. Take the denominator and draw in crossing arrows. Thus:

$$\begin{vmatrix} 5 & 2 \\ 3 & -1 \end{vmatrix}$$

Step 3. Starting at the upper left (5) and going to the lower right (-1), form the product of (-5).

Step 4. Starting at the lower left (3) and going to the upper right (2), form the product of (6); however, multiplication in this direction always requires that the product be multiplied by a (-1). Therefore, the product is (-6).

Step 5. Algebraically add the results of Steps 3 and 4 to obtain (-5) + (-6) = -11.

Step 6. Repeat Steps 2 through 5 for the numerator of the determinant array. Thus:

$$I_1 = \frac{\begin{vmatrix} 16 & 2 \\ 3 & -1 \end{vmatrix}}{-11} = \frac{(-16) + (-6)}{-11} = \frac{-22}{-11} = 2A$$

$$I_2 = \frac{\begin{vmatrix} 5 & 16 \\ 3 & 3 \end{vmatrix}}{-11} = \frac{(+15) + (-48)}{-11} = \frac{-33}{-11} = 3A$$

NOTE: The denominator remains the same throughout the problem.

172

EXAMPLE 2:

A certain network was analyzed using nodal equations The resulting equations were as follows:

$$-6V_1 = -18 + 8V_2 \quad \text{and} \quad 10 = 4V_2 + 8V_1$$

Determine V_1 and V_2.

SOLUTION: Follow the steps of the preceding example.

$$K_1 = G_{11}V_1 + G_{12}V_2$$

$$K_2 = G_{21}V_1 + G_{22}V_2$$

Place equations in standard form.

$$18 = 6V_1 + 8V_2$$

$$10 = 8V_1 + 4V_2$$

Substitute into the determinant array. Thus:

$$V_1 = \frac{\begin{vmatrix} 18 & 8 \\ 10 & 4 \end{vmatrix}}{\begin{vmatrix} 6 & 8 \\ 8 & 4 \end{vmatrix}} = \frac{72 - 80}{24 - 64} = \frac{-8}{-40} = 200 \text{ mV}$$

$$V_2 = \frac{\begin{vmatrix} 6 & 18 \\ 8 & 10 \end{vmatrix}}{-40} = \frac{60 - 144}{-40} = \frac{-84}{-40} = 2.1 \text{ V}$$

EXAMPLE 3:

In order to determine the value of current I_1 in a circuit, the following equations were developed.

$$1I_3 = -11 + 4I_1 + 6I_2$$

$$5 = 10I_1 + 2I_2 - 4I_3$$

$$-3I_1 = 6I_3 - 4I_2 - 9$$

SOLUTION:

Step 1. Place the equations in standard form.

$$K_1 = R_{11}I_1 + R_{12}I_2 + R_{13}I_3$$

$$K_2 = R_{21}I_1 + R_{22}I_2 + R_{23}I_3$$

$$K_3 = R_{31}I_1 + R_{32}I_2 + R_{33}I_3$$

$$11 = 4I_1 + 6I_2 - 1I_3$$

$$5 = 10I_1 + 2I_2 - 4I_3$$

$$9 = 3I_1 - 4I_2 + 6I_3$$

Step 2. Set up a third order determinant array.

$$
I_1 = \frac{\begin{vmatrix} K_1 & R_{12} & R_{13} \\ K_2 & R_{22} & R_{23} \\ K_3 & R_{32} & R_{33} \end{vmatrix} \begin{matrix} K_1 & R_{12} \\ K_2 & R_{22} \\ K_3 & R_{32} \end{matrix}}{\begin{vmatrix} R_{11} & R_{12} & R_{13} \\ R_{21} & R_{22} & R_{23} \\ R_{31} & R_{32} & R_{33} \end{vmatrix} \begin{matrix} R_{11} & R_{12} \\ R_{21} & R_{22} \\ R_{31} & R_{32} \end{matrix}}
$$

Notice how Columns 1 and 2 are repeated in both the numerator and denominator.

Step 3. Take the denominator of the determinant and draw in the arrows.

$$
\begin{vmatrix} R_{11} & R_{12} & R_{13} \\ R_{21} & R_{22} & R_{23} \\ R_{31} & R_{32} & R_{33} \end{vmatrix} \begin{matrix} R_{11} & R_{12} \\ R_{21} & R_{22} \\ R_{31} & R_{32} \end{matrix}
$$

$$
\begin{vmatrix} 4 & 6 & -1 \\ 10 & 2 & -4 \\ 3 & -4 & 6 \end{vmatrix} \begin{matrix} 4 & 6 \\ 10 & 2 \\ 3 & -4 \end{matrix}
$$

Step 4. Starting at the upper left, 1st column, and proceeding to the lower right, 3rd column:

$4 \times 2 \times 6 = 48$

Start at the top of Column 2 and proceed diagonally downward (include 3 columns):

$6 \times (-4) \times 3 = -72$

Start at the top of Column 3 and proceed diagonally downward (include 3 columns):

$(-1) \times 10 \times (-4) = 40$

Adding these terms algebraically:

$48 + (-72) + 40 = 16$

Step 5. Repeat Step 4 going diagonally upward. Multiply each set of diagonal arrows by a (−1).*

$3 \times 2 \times (-1) \times (-1)^* = 6$

$(-4) \times (-4) \times 4 \times (-1)^* = -64$

$6 \times 10 \times 6 \times (-1)^* = -360$

Adding these terms algebraically:

$$6 + (-64) + (-360) = -418$$

Step 6. Add the results of Steps 4 and 5 algebraically.

$$16 + (-418) = -402 \quad \text{(denominator reduction)}$$

Step 7. Repeat Steps 2 through 6 for the numerator of the determinant.

$$
\begin{vmatrix}
11 & 6 & -1 \\
5 & 2 & -4 \\
9 & -4 & 6
\end{vmatrix}
\begin{matrix}
11 & 6 \\
5 & 2 \\
9 & -4
\end{matrix}
$$

$$[11 \times 2 \times 6] + [6 \times (-4) \times 9] + [(-1) \times 5 \times (-4)] = 132 + (-216) + 20 = -64$$

$$
\begin{vmatrix}
11 & 6 & -1 \\
5 & 2 & -4 \\
9 & -4 & 6
\end{vmatrix}
\begin{matrix}
11 & 6 \\
5 & 2 \\
9 & -4
\end{matrix}
$$

$$[9 \times 2 \times (-1) \times (-1)^*] + [(-4) \times (-4) \times 11 \times (-1)^*] + [6 \times 5 \times 6 \times (-1)^*]$$

$$= 18 + (-176) + (-180) = -338$$

Therefore: $(-64) + (-338) = -402$ (numerator reduction)

Step 8.

$$I_1 = \frac{-402}{-402} = 1 \text{ A}$$

It might be well to determine I_2 and I_3 to make certain that you have command of the determinant process.

$$I_2 = 1.5 \text{ A} \quad \text{and} \quad I_3 = 2 \text{ A}$$

D-1 PREPARING THE LABORATORY REPORT

The following topics are included to assist in the preparation of the laboratory report. The suggested format is provided only as a guide. Consult your instructor for specific instructions.

TITLE PAGE

The title page is a cover sheet which includes your name, the date that the experiment was performed, the class (section), and the experiment number and title. Some form of grading sheet may also be included by your instructor.

SCHEMATIC

A schematic of each circuit constructed during the experiment should be included. It should be neatly drawn (using a straight edge and a template) on quadrille paper. All values should be noted and all symbols clearly identified and labelled. In addition, the title of the experiment should appear on the schematic.

LIST OF MATERIALS

The data tables should be neat, accurate, and complete. In an effort to make the data tables as attractive as possible, record your rough data in a notebooks or on quadrille paper, and then transfer it to the data tables. Whenever possible all graphs should appear on graph paper. The graph must be accurately plotted, labelled and properly identified.

DATA INTERPRETATION AND CONCLUSIONS

A short, concise paragraph should be written discussing the points brought out in the individual "Data Interpretation and Conclusions" sections.

PROBLEMS

The problems given at the end of the experiment should be answered on a separate sheet of paper in a complete and concise manner.

JOE SMITH
MARCH 18, 1982
ELECTRO 41

EXPERIMENT   3
AC VOLTAGE MEASUREMENTS

AC VOLTAGE MEASUREMENT
EXPERIMENT 3
SCHEMATICS

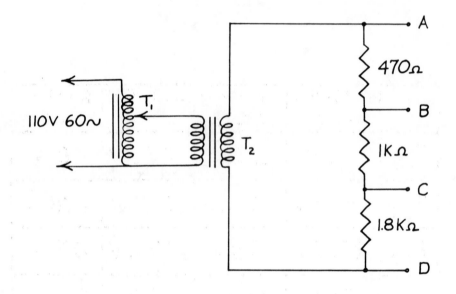

MATERIAL LIST

AC VTVM — SIMPSON, MODEL #715, SER. #8826
OSCILLOSCOPE— EICO, MODEL #460, SER. #12A449
(T₁) VARIABLE TRANSFORMER—
        SUPERIOR ELECTRIC CO.
        MODEL #116
(T₂) ISOLATION TRANSFORMER—
        TRIAD, MODEL #N-53M
RESISTORS — 470Ω, 1KΩ, 1.8KΩ
CLIP LEADS

JOE SMITH
MARCH 18, 1982
ELECTRO 41

EXPERIMENT 3 – DATA

	MEASURED RMS	P–P VOLTS	CALCULATED RMS	% DIFF. RMS
E_{AB}	7.1 V	20.2 V	7.25 V	2.8 %
E_{BC}	15.15 V	43.5 V	15.35 V	1.38 %
E_{CD}	27.7 V	78 V	27.5 V	0.7 %
E_{AD}	50 V	142 V	50.2 V	0.4 %

DATA INTERPRETATION AND CONCLUSION

THE AVERAGE DIFFERENCE OF 1.31%
BETWEEN THE VOLTAGE MEASURED
WITH THE AC VTVM AND THE VOLTAGE
MEASURED WITH THE SCOPE CAN BE
ATTRIBUTED TO THE EQUIPMENT
MANUFACTURER'S TOLERANCE. SINCE
THE CALIBRATING VOLTAGE OF THE
SCOPE HAS A TOLERANCE OF 3%
AND THE AC VTVM HAS A TOLERANCE
OF 6% (IN THE AREA WHERE THE
VOLTAGES WERE READ), THE ERROR
IN THE EXPERIMENT IS DUE TO THE
INHERENT ERROR OF THE EQUIPMENT.

IN CONCLUSION, THE AC VTVM,
WHEN READ IN THE UPPER 2/3 OF
THE SCALE, WILL GIVE A READING
WITHIN 10%, WHILE THE OSCIL-
LOSCOPE, WHEN ACCURATELY CAL-
IBRATED, WILL HAVE EVEN BETTER
ACCURACY. THE METER INDICATES
RMS VOLTAGE WHILE THE SCOPE
INDICATES PEAK TO PEAK VOLTAGE.

JOE SMITH
MARCH 18, 1982
ELECTRO 41

PROBLEMS:

1. WHEN THE SCOPE READS 64 V. P-P THE AC VTVM WOULD READ 22.6 V. .

2. WHEN THE AC VTVM READS 51.3 V. RMS THE SCOPE WOULD READ 145 V. P-P .

D-2 REFERENCES

1. Angerbauer, George. **Principles of DC and AC Circuits.** Massachusetts, Duxbury Press, 1977.

2. Angus, Robert Brownell. **Electrical Engineering Fundamentals.** 2nd Edition. Massachusetts, Addison-Wesley Publishing Company, Inc., 1968.

3. Alvarez, E.C. and Tontsch, John. **Fundamental Circuit Analysis.** Chicago, Science Research Associates, Inc., 1978.

4. Boylestad, Robert L. **Introductory Circuit Analysis.** 4th Edition. Ohio, Charles E. Merrill Publishing Company, 1982.

5. DeFrance, J.J. **Electrical Fundamentals.** 2nd Edition. New Jersey, Prentice-Hall, 1969.

6. Gillie, Angelo C. **Electrical Principles of Electronics.** Rev. Edition. New York, McGraw-Hill Book Company, 1977.

7. Grob, Bernard. **Basic Electronics.** 4th Edition. New York, McGraw-Hill Book Company, 1977.

8. Harter, James and Lin, Paul. **Essentials of Electric Circuits.** Virginia, Reston Publishing Company, Inc., 1982.

9. Hickey, Henry V. and Villines, William M. Jr. **Elements of Electronics.** 4th Edition. New York, McGraw-Hill Book Company, 1090.

10. Jackson, Herbert W. **Introduction to Electric Circuits.** 5th Edition. New Jersey, Prentice-Hall, Inc., 1981.

11. Key, Eugene George. **Principles of Electricity.** New York, Harper and Row, 1967.

12. Lurch, Norman E. **Electric Circuit Fundamentals.** New Jersey, Prentice-Hall, Inc., 1979.

13. Mottershead, Allen. **Introduction to Electricity and Electronics.** New York, John Wiley and Sons, Inc., 1982.

14. Oppenheimer, Samuel L. and Borchers, Jean Paul. **Direct and Alternating Currents.** 2nd Edition. New York, McGraw-Hill Book Company, 1973.

15. Romanowitz, H. Alex. **Introduction to Electronics.** 2nd Edition. New York, John Wiley and Sons, Inc., 1976.

16. Shrader, Robert L. **Electrical Fundamentals for Technicians.** 2nd Edition. New York, McGraw-Hill Book Company, 1977.

17. Slurzburg, Morris and Osterheld, William. **Essentials of Electricity-Electronics.** 3rd Edition. New York, McGraw-Hill Book Company, 1965.

18. Veatch, Henry C. **Electrical Circuit Action.** Chicago, Science Research Associates, Inc., 1978.

1 - METER FUNDAMENTALS
DATA TABLES

Name _____

Date _____

SECTION A: LINEAR SCALES

Class _____

List Mfg. _____ and Model No. _____

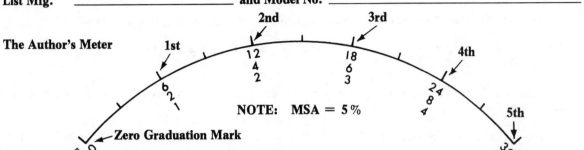

The Author's Meter

NOTE: MSA = 5%

FIGURE 1-12

	GUIDE (Using Figure 1-12)				EXPERIMENT (Answers using your meter)		
1.	DC			1.			
2.	0-30 Top	0-10 Middle	0-5 Lower	2.			
3.	5			3.			
4.	2			4.			
5.	6 Top	2 Middle	1 Lower	5.			
6.	3	1	0.5	6.			
7.	18	6	3	7.			
8.	15	5	2.5	8.			

9.   (The authors did this in Figure 1-12)   9.

See Figure 1-12

10.	± 5%	10.	
11.	20%	11.	

SECTION B: NON-LINEAR SCALES

Name _____

Date _____

Class _____

1. _____

2. _____

3. _____

6.

4. _____

5. _____

7. _____

8. _____

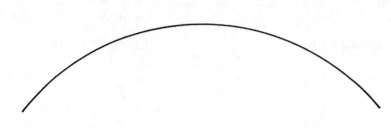

SECTION C: RANGE SWITCH

1(a) _____   1(b) _____

2(a)   2(b)

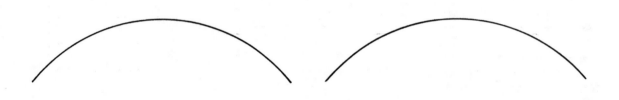

1(c) _____   1(d) _____

2(c)   2(d)

3. _____

2 - RESISTIVE DEVICES AND PARAMETERS DATA TABLES

Name _____

Date _____

Class _____

SECTION A

7. a. _____ b. _____ c. _____ d. _____

8. Zero: Right _____ Left _____

9.

	(a) RANGE SETTING	(b) MEASURED VALUE	(c) STAMPED VALUE
1			
2			
3			
4			
5			
6			

SECTION B

	(1) MEASURED VALUE	(2a) COLOR CODE VALUE	% TOL.	2(b) WITHIN TOL.	2(c) % DIFF.
1					
2					
3					
4					
5					
6					
7					
8					
9					
10					
11					
12					

EXPERIMENT 2

SECTION C

1.

	BANDED END	**UNBANDED END**
Brown	has continuity with	_____
Red	has continuity with	_____
Orange	has continuity with	_____
Yellow	has continuity with	_____
Green	has continuity with	_____
Blue	has continuity with	_____
Violet	has continuity with	_____
Gray	has continuity with	_____

2. a. Resistance between orange and blue _____

   b. Estimated Distance —————————————

   c. Distance —————————————

3. Draw the schematic of Test Cable No. 2.

   Brown

   Red

   Orange

   Yellow

   Green

   Blue

   Violet

   Gray

3 - BATTERIES AND PHOTOVOLTAIC CELLS:
DATA TABLES

Name _____

Date _____

Class _____

SECTION A:

6. _____

SECTION B

1. _____   2. _____   3. _____

4. _____   5. _____   6. _____

SECTION C

1.

R_{int} (ohms)	ELAPSED TIME (hours)	E_0 (volts)	R_{int} (ohms)	ELAPSED TIME (hours)	E_0 (volts)
4.7 Ω	1.0		2.2 kΩ	5.5	
10 Ω	2.0		2.7 kΩ	6.0	
100 Ω	3.0		3.9 kΩ	6.5	
470 Ω	4.0		4.7 kΩ	7.0	
1 kΩ	4.5		5.6 kΩ	7.5	
1.5 kΩ	5.0		6.8 kΩ	8.0	

4. The end point voltage occurs at time equal to _____

SECTION D

2. E_{ba} _____ E_{bc} _____ E_{ac} _____

4. E_{ab} _____ E_{cb} _____ E_{ca} _____

5. E_{ca} _____ E_{bc} _____ E_{ba} _____

   E_{ab} _____ E_{ac} _____ E_{cb} _____

SECTION E

6. Yes _____   No _____

EXPERIMENT 3

4 - SERIES CIRCUIT ANALYSIS
DATA TABLES

Name _____

Date _____

Class _____

SECTION A

4. _____   5. _____

SECTION B

3 E_{ac} _____   4. $R_T = R_{ac}$ _____

5. $E_{supply} = E_{ac}$ _____   6. % Differences _____

SECTION C

2. and 3.   Make a table of values for voltage and current.

8. a _____   b _____   c _____   d _____

SECTION D

1. R_T_____   4. $E_{source} = E_{ba}$ _____

7. R_T _____ $E_{source} = E_{ba}$ _____

8. % Difference of source voltage _____

   % Difference of total current _____

   % Difference of total resistance _____

9. Yes _____   No _____

SECTION E

2. E_{ba} _____ E_{cb} _____ E_{dc} _____ E_{ed} _____

   E_{ae} _____

3. $E_{ba} (\pm) E_{cb} (\pm) E_{dc} (\pm) E_{ed} (\pm) E_{ae} = 0$

   _____ _____ _____ _____ _____ = _____

   _____ ≈ 0

4. Write an equation for clockwise movement around the closed loop starting at Point **a**. (Supply the voltage subscripts and direction signs.)

   E _____   E _____   E _____   E _____   E _____

   _____   _____   _____   _____   _____

   _____ = 0

EXPERIMENT 4

SECTION F

2. Supply the voltage subscripts and direction signs.

   E _____ E _____ E _____ E _____ E _____ E _____ E _____ E _____

3. Place the equation here.

4.

SECTION G

3. E_s _____ E_o _____ R_L _____ R_{int} _____

4. % difference _____

5. E_s _____ E_o _____ R_L _____

   R_{int} _____ % Difference _____

5 - PARALLEL CIRCUIT ANALYSIS
DATA TABLES

Name _____

Date _____

Class _____

SECTION A

2. I_T _____   3. I_T _____

4. I_T _____

5. The total circuit current increased _____

_____

6. I_1 without lamp B _____ I_1 without lamps C or B _____

Brightness changed?   Current changed?

Yes _____ No _____   Yes _____ No _____

7. E_{ba} _____ E_{dc} _____ E_{fe} _____ E_{hg} _____

8. E_{ha} _____ E_{bg} _____

_____

9. Voltage changed?   Yes _____   No _____

10. _____

_____

11. Yes _____ No _____ Why? _____

_____

SECTION B

1. I_T _____ I_1 _____ I_2 _____ I_3 _____

2. I_T _____

3. I_1 _____ I_2 _____ I_3 _____

4.   =

5. I_T _____ I_1 _____ I_2 _____ I_3 _____

=

6. % Difference _____

SECTION C

1. I_T (meas.) _____   2. I_1 (cal.) _____

3. I_1 (meas.) _____   4. I_2 (cal.) _____ I_2 (meas.) _____

5. $I_T = I_1 + I_2$ (cal.) _____ $I_T = I_1 + I_2$ (meas.) _____

   % Difference _____

SECTION D

2. E_{ba} _____ E_{cb} _____ E_{dc} _____

   E_{ed} _____ E_{fe} _____ E_{af} _____

3. Supply the subscript and polarity

   E _____ E _____ E _____ E _____ E _____ E _____ = 0

   _____ _____ _____ _____ _____ _____ = 0

   _____ = 0

4.

6 - COMPOUND CIRCUIT ANALYSIS
DATA TABLES

Name _____

Date _____

Class _____

SECTION A

1.

3. R_1 series _____ parallel _____ R_2 series _____ parallel _____

   R_3 series _____ parallel _____ R_4 series _____ parallel _____

4.

R_1 series _____ parallel _____ R_2 series _____ parallel _____

R_3 series _____ parallel _____ R_4 series _____ parallel _____

R_5 series _____ parallel _____

EXPERIMENT 6

SECTION B

2. I_T _____

3. I_T _____ Value of the combined resistance of R_3 and R_4 _____

4.

5. I_T _____ Value of the combined resistance of R_2 and R_{eq-1} _____

6.

7. I_T _____ Value of R_{eq} _____

8. Yes _____ No _____ Why? _____

_____

9.

SECTION C

3. R_T _____   Instructor's OK _____

4.

SECTION D

1. E_{ba} _____ E_{cb} _____ E_{dc} _____

2. E_{ba} _____ E_{cb} _____ E_{dc} _____

3. R point **a** to **b** _____   R point **b** to **c** _____   R point **c** to **d** _____

4. R point **a** to **b** _____   R point **b** to **c** _____   R point **c** to **d** _____

5.

SECTION E

2. E_{ba} _____ E_{gb} _____ E_{hg} _____ E_{ah} _____

3.

SECTION E - Continued

4.

5.

SECTION F

1. Current entering:

   Node a _____   _____   _____

   Node b _____

   Node c _____

   Current leaving:

   Node a _____

   Node b _____   _____

   Node c _____   _____   _____

   $I_{entering} = I_{leaving}$

   Node a: _____ + _____ + _____ = _____

   _____ = _____

   Node b: _____ = _____ + _____

   _____ = _____

   Node c: _____ = _____ + _____ + _____

   _____ = _____

2. % Difference:

   Node a _____   Node b _____   Node c _____

7 - METER MULTIPLIERS AND SHUNTS
VOM LOADING EFFECTS
DATA TABLES

Name _____

Date _____

Class _____

SECTION A

4. R_{meq} _____   5. _____   6. R_{meq}(cal.) _____

7. % Difference _____

8. R_{meq}(meas.) _____   Current Spec. _____

   R_{meq}(cal.) _____   % Difference _____

SECTION B

1. Current Calculated

   Node a _____

   Node b _____

   Node c _____

2. Current Measured

   Node a _____

   Node b _____

   Node c _____

3. % Difference (Steps 1 and 2)

   Node a _____

   Node b _____

   Node c _____

4. (a)

4. (b)

SECTION B - Continued

4. (c)

5. Recalculated Currents   6. % Difference (Steps 2 and 5)

Node a _____   Node a _____

Node b _____   Node b _____

Node c _____   Node c _____

SECTION C

1. I_T(cal.) _____   3. R_{shunt} _____   Instructor's OK _____

5. I_T(meas.) _____   6. % Difference _____

7. I_T(cal.) _____ R_{shunt} _____   Instructor's OK _____

   I_T(meas.) _____   % Difference _____

SECTION D

5. R_{meq}(meas.) _____   6. R_{meq}(cal.) _____   7. %Diff. _____

8. R_{meq}(meas.) _____ R_{meq}(cal.) _____   % Diff. _____

SECTION E

Name _____

Date _____

Class _____

1. $E_{100 \text{ k}\Omega}$(cal.) _____

2. $E_{100 \text{ k}\Omega}$(meas.) _____

3. % Diff. _____

4.

5. $E_{100 \text{ k}\Omega}$(recal.) _____

6. % Difference (Steps 2 and 5) _____

7. $E_{100 \text{ k}\Omega}$(cal.) _____ $E_{100 \text{ k}\Omega}$(meas.) _____ % Diff. _____

   Schematic

   $E_{100 \text{ k}\Omega}$(recal.) _____ % Difference _____

8. $E_{100 \text{ k}\Omega}$(cal.) _____ $E_{100 \text{ k}\Omega}$(meas.) _____ % Diff. _____

   Schematic

   $E_{100 \text{ k}\Omega}$(recal.) _____ % Difference _____

EXPERIMENT 7

SECTION F

1. E_{R-1}(cal.) _____ E_{R-2}(cal.) _____ 2. R_{mult} _____

4. E_{R-1}(meas.) _____ E_{R-2}(meas.) _____ 5. % Diff. _____

6. Sensitivity _____

7. E_{R-1}(cal.) _____ R_{mult} _____ E_{R-1}(meas.) _____

   % Difference _____ Sensitivity _____

8. E_{R-1}(cal.) _____ R_{mult} _____ E_{R-1}(meas.) _____

   % Difference _____ Sensitivity _____

8 - INTRODUCTION TO POWERED MULTIFUNCTION METERS DATA TABLES

Name _____

Date _____

Class _____

SECTION A

2.

	COLOR CODE VALUE	MEASURED VALUE	% DIFFERENCE
1			
2			
3			
4			
5			
6			

SECTION B

2. E_{de} _____ E_{ba} _____ E_{ab} _____ E_{cd} _____

E_{cb} _____ E_{ed} _____ E_{bd} _____ E_{db} _____

E_{bc} _____ E_{dc} _____

SECTION C

1. Current into node D _____

2. Current into node B _____   Current into node C _____

3. R_{ae} _____ I_T _____

4. % Difference between I_T and the current into node D _____

5.

SECTION D

2. VOM voltage (R × 1) _____   Black lead + _____ − _____

3. VOM voltage:

    R × 10 _____   R × 10 k _____

    R × 100 _____   R × 100 k _____

    R × 1 k _____   R × 1 M _____

SECTION E

1. E_1 _____

2. E_2 _____

3. R_{meq} _____

9 - VOLTAGE DIVIDERS
DATA TABLES

Name _____

Date _____

Class _____

SECTION A

1. R_{ac} _____   Yes _____   No _____

2. R_{ba} (ccw) _____ R_{ba} (cw) _____   Yes _____   No _____

3. R_{bc}(ccw) _____ R_{bc}(cw) _____   Yes _____   No _____

SECTION B

1.

E_{bc}	E_{ab}	R_{ba}	R_{bc}	E_{bc}	E_{ab}	R_{ba}	R_{bc}
0				6			
1				7			
2				8			
3				9			
4				10			
5							

2. $\dfrac{2}{\quad}$ = ___ = $\dfrac{5}{\quad}$ = ___ = $\dfrac{8}{\quad}$ = ___ =

3. $\dfrac{3}{\quad}$ = ___ = $\dfrac{6}{\quad}$ = ___ = $\dfrac{9}{\quad}$ = ___ =

SECTION C

1. R_1 _____ R_2 _____

2. E_o _____ to _____

SECTION D

1.

EXPERIMENT 9

SECTION D - Continued

2. R_1 _____ R_2 _____

3. $P_1 \times 3$ _____ $P_2 \times 3$ _____

4. $P_{pot} \times 2$ _____

5. R_1 _____ R_2 _____

10 - SUPERPOSITION THEOREM
DATA TABLES

Name _____

Date _____

Class _____

SECTION A

1. I_1 _____ I_2 _____ I_3 _____
2.

5.

6.

7. I_1 _____ I_2 _____ I_3 _____
8. I_1 _____ I_2 _____ I_3 _____

SECTION A - Continued

9. % diff. I_1 _____ % diff. I_2 _____ % diff. I_3 _____

10. % diff. I_1 _____ % diff. I_2 _____ % diff. I_3 _____

SECTION B

1. I_{100} _____ I_{330} _____ I_{150} _____ I_{56} _____ I_{47} _____

2.

3. I_{100} _____ I_{330} _____ I_{150} _____ I_{56} _____ I_{47} _____

5. I_{100} _____ I_{330} _____ I_{150} _____ I_{56} _____ I_{47} _____

6. I_{100} _____ I_{330} _____ I_{150} _____ I_{56} _____ I_{47} _____

SECTION B - Continued

7.

8. I_{100} _____ I_{330} _____ I_{150} _____ I_{56} _____ I_{47} _____

9. I_{100} _____ I_{330} _____ I_{150} _____ I_{56} _____ I_{47} _____

10. % diff. I_{100} _____   % diff. I_{330} _____   % diff. I_{150} _____

   % diff. I_{56} _____   % diff. I_{47} _____

11. % diff. I_{100} _____   % diff. I_{330} _____   % diff. I_{150} _____

   % diff. I_{56} _____   % diff. I_{47} _____

11 - THEVENIN'S THEOREM
DATA TABLES

Name _____

Date _____

Class _____

SECTION A

3. E_{ba} (a) _____ E_{ba} (b) _____

4. I_{ba} (a) _____ I_{ba} (b) _____

5. R_{ba} (a) _____ R_{ba} (b) _____

6. % diff. E _____   % diff. I _____   % diff. R _____

SECTION B

1. E_{th} (cal.) _____ R_{th} (cal.) _____ E_o (cal.) _____

2. E_{th} (meas.) _____ R_{th} (meas.) _____ E_o (meas.) _____

3. E_o (meas.) _____

4. % diff. E_o _____

SECTION C

2. E_{th} _____ E_o _____   3. R_{th} _____

4.

EXPERIMENT 11

SECTION D

1. E_{th} _____ R_{th} _____

2.

3. P_L _____

12 - MAXIMUM POWER TRANSFER
DATA TABLES

Name _____

Date _____

Class _____

SECTION A

2. E_{ba} for:

1 kΩ _____   3.3 kΩ _____   6.8 kΩ _____   10 kΩ _____

12 kΩ _____   15 kΩ _____   27 kΩ _____   39 kΩ _____

3. P_L for:

1 kΩ _____   3.3 kΩ _____   6.8 kΩ _____   10 kΩ _____

12 kΩ _____   15 kΩ _____   27 kΩ _____   39 kΩ _____

4. Place table here

6. R at max. power _____

SECTION B

1. R_{th} _____ E_{th} _____

2.

3. % diff. _____

EXPERIMENT 12

SECTION C

2. $E_{unloaded}$ _____ E_{loaded} _____

3. R_{int} _____

4. R_L at max. power _____

5. % diff. _____

6.

SECTION D

1. I_T _____

2. I_L _____   3. I_{R2} _____

4. R_1 _____ R_2 _____

   0 =

5. R_3 _____

7. R_{th} _____ E_L _____

13 - NORTON'S THEOREM
DATA TABLES

Name _____

Date _____

Class _____

SECTION A

1. I_{sc} _____ R_{th} _____

2. I_L (100 Ω) _____ I_L (220 Ω) _____ I_L (470 Ω) _____ I_L (1 kΩ) _____

4. I_L (100 Ω) _____ I_L (220 Ω) _____ I_L (470 Ω) _____ I_L (1 kΩ) _____

5. % diff. I_L (100 Ω) _____   % diff. I_L (220 Ω) _____

    % diff. I_L (470 Ω) _____   % diff. I_L (1 kΩ) _____

6. Yes _____ No _____ _____ mA

SECTION B

3. I_L (10 Ω) _____ I_L (47 Ω) _____ I_L (100 Ω) _____

4. Yes _____ No _____

SECTION C

2. I_L (47 Ω) _____ I_L (100 Ω) _____ I_L (470 Ω) _____

    I_L (1 kΩ) _____ I_L (2.7 kΩ) _____ I_L (4.7 kΩ) _____

    I_L (10 kΩ) _____

SECTION D

2. $E_{470 Ω}$ _____ $E_{1 kΩ}$ _____ $E_{10 kΩ}$ _____ $E_{1 MΩ}$ _____

3. Yes _____ No _____

14 - NON-LINEAR DEVICES
DATA TABLES

Name _____

Date _____

Class _____

SECTION A

2. Place table here

4. Yes_____   No _____

SECTION B

3. Place table here

5. R_{cell} at 1 mA _____   6. R_{cell} at 0.1 mA _____

7. Yes_____   No _____

8. Place table here

10. Yes_____   No _____

SECTION C

2. Place table here

3. Place table here

8. R at 0.4 V _____   9. R at 40 V _____

11. r _____

SECTION D

3. Place table here

5. R for 1 mA _____

6. R for 35 mA _____

7. r _____

SECTION E

3. Diode _____   Photoconductive cell _____

Lamp _____   Thermistor _____

15 - MILLMAN'S THEOREM
DATA TABLES

Name _____

Date _____

Class _____

SECTION A

1. E_{ab}(cal.) _____

2. E_{ab}(meas.) _____

3. % diff. E_{ab} _____

SECTION B

1. R of thermistor _____

2.

3. E_{ag} (diode conducting) _____ E_{ag} (diode not conducting) _____

4. Yes_____   No _____

5. E_{ag} (diode conducting) _____ E_{ag}(diode not conducting) _____

6. Yes_____   No _____

7. E_{ag} (meas.) _____

8. % diff. E_{ag} _____

9. E_{ag} (meas.) _____

10. % diff. E_{ag} _____

11. Record your observations.

EXPERIMENT 15

SECTION C

1.

2. E_{ag} (dark) _____ E_{ag} (light) _____

4. E_{ag} (meas.) _____

5. % diff. E_{ag} _____

6. E_{ag}(meas. light) _____

7. % diff. E_{ag} _____

16 - LOOPS AND NODES
DATA TABLES

Name _____

Date _____

Class _____

SECTION A

1. E_{ba} (cal.) _____

2. E_{ba} (meas.) _____

3. % diff. E_{ba} _____

SECTION B

1. $I_{3.3\ k\Omega}$ (cal.) _____

2. $I_{3.3\ k\Omega}$ (meas.) _____

3. % diff. $I_{3.3\ k\Omega}$ _____

SECTION C

1. $I_{3.3\ k\Omega}$ (cal.) _____

2. $I_{3.3\ k\Omega}$ (meas.) _____

3. % diff. $I_{3.3\ k\Omega}$ _____

17 - INDUCTANCE
DATA TABLES

Name _____

Date _____

Class _____

SECTION A

1. NE-2 fires at _____

2. R_{high} _____ R_{low} _____

4. _____

_____

_____

SECTION B

3. _____

_____

_____

SECTION C

3. _____

_____

_____

SECTION D

4. _____

_____

_____

6. _____

_____

_____

18 - CAPACITANCE DATA TABLES

Name _____

Date _____

Class _____

SECTION A

1. Q (cal.) _____   3. E (meas.) _____

4. E (cal.) _____   5. % diff. _____

6. Q (cal.) _____   7. % diff. _____

SECTION B

1. $E_{.1}$ _____ $E_{.5}$ _____

2. $Q_{.1}$ _____ $Q_{.5}$ _____

3. Q_T _____

4. % diff. $Q_{.1}$ and Q_T _____   % diff. $Q_{.5}$ and Q_T _____

SECTION C

1. E_{C1} _____ E_{C2} _____   Yes _____ No _____

2. E_{C1} _____ E_{C2} _____   Yes _____ No _____

3. % diff. _____

SECTION D

3. I_T _____   4. R_p _____

5.

EXPERIMENT 18

19 - SWITCHES—CHARACTERISTICS
AND RATINGS
DATA TABLES

Name _____

Date _____

Class _____

SECTION B

Instructor's OK _____

Schematic:

20 - RELAY CHARACTERISTICS
DATA TABLES

Name _____

Date _____

Class _____

SECTION A

2. Pull-in current _____

3. Drop-out current _____

4. Average pull-in current _____   Average drop-out current _____

5. Pull-in voltage _____

6. Drop-out voltage _____

7. Pull-in voltage _____   Drop-out voltage _____

8. % diff. pull-in voltage _____   % diff. drop-out voltage _____

SECTION B

1. E_{CL704L}(cal.) _____

2. E_{CL704L}(meas.) _____

   Pull-in current _____

3. % diff. _____

4. % diff. pull-in current _____

6. Pull-in time _____

7. Pull-in time _____

SECTION C

1. Relay _____ closes E_{ag} _____

2. Relay _____ closes E_{ag} _____

3. Relay _____ closes E_{ag} _____

5. E_{ag} _____   % diff. E_{ag} _____   Relay _____ closes

6. E_{ag} _____   % diff. E_{ag} _____   Relay _____ closes

7. E_{ag} _____   % diff. E_{ag} _____   Relay _____ closes

21 - RC TIME CONSTANTS
DATA TABLES

Name _____

Date _____

Class _____

SECTION A

1. $E_{pull-in}$ _____ $E_{drop-out}$ _____

2. I_{leak} _____ R_p _____

3. Pull-in time (cal.) _____   4. Pull-in time (meas.) _____

5. % diff. in pull-in times _____

6. Drop-out time (cal.) _____   7. Drop-out time (meas.) _____

8. % diff. in drop-out times _____

SECTION B

1. Prediction _____

_____

2. Yes _____   No _____

3. Prediction _____

_____

_____

Prediction correct?   Yes _____   No _____

4. Time lapse (cal.) _____ sec.   5. Time lapse (meas.) _____ sec.

6. % diff. _____

7. Cycle time _____ sec.

EXPERIMENT 21

22 - BRIDGE CIRCUITRY
DATA TABLES

Name _____

Date _____

Class _____

SECTION A

3. R_{pot} _____ R_x (cal.) _____

4. R_x (meas.) _____   5.   % diff. R_x _____

6. R_{pot} _____ R_x (cal.) _____ R_x (meas.) _____

    % diff. R_x _____

SECTION B

1. R_{pot} _____   2. E_d (cal.) _____

3. E_d (meas.) _____   4.   % diff. E_d _____

SECTION C

1. $I_{2.7 \ k\Omega}$ (cal.) _____   2. $I_{2.7 \ k\Omega}$ (meas.) _____

3. % diff. $I_{2.7 \ k\Omega}$ _____

SECTION D

2. _____

    _____

    _____

3. _____

    _____

    _____

4. _____

    _____

    _____

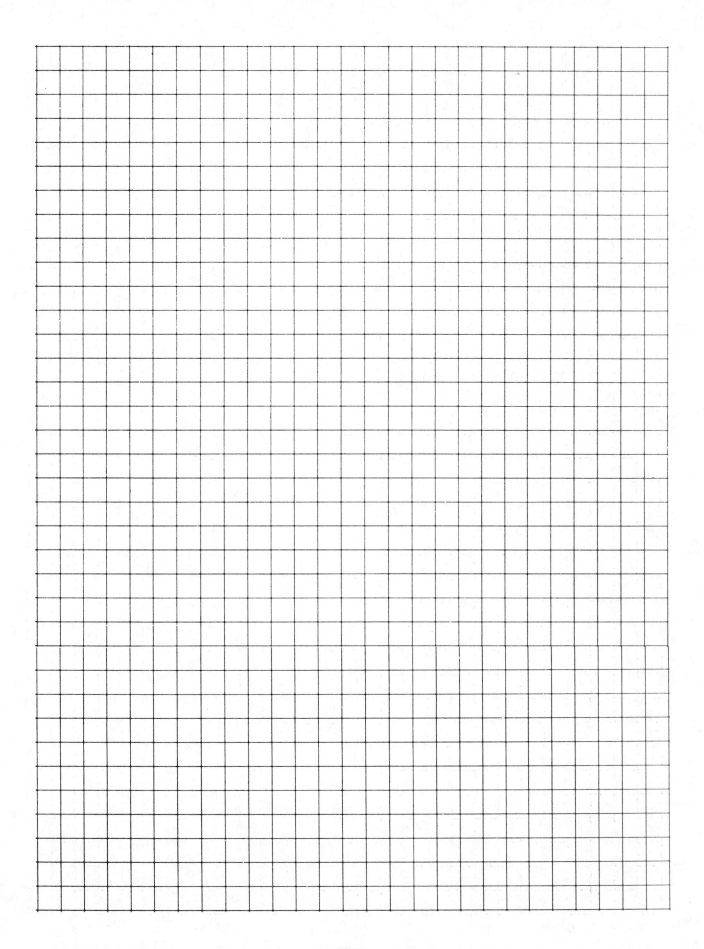

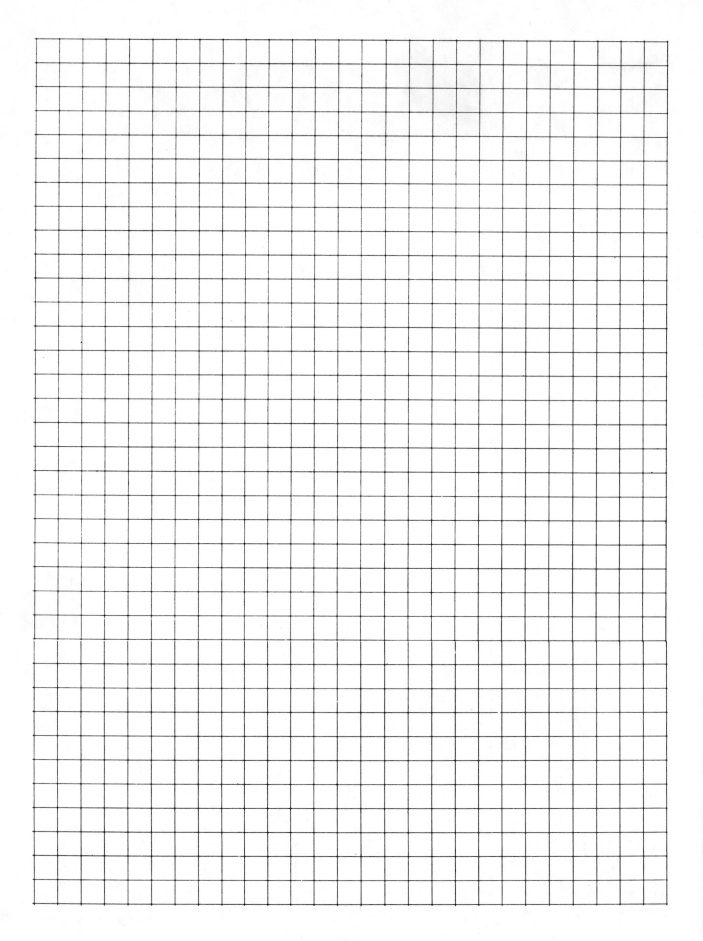

236

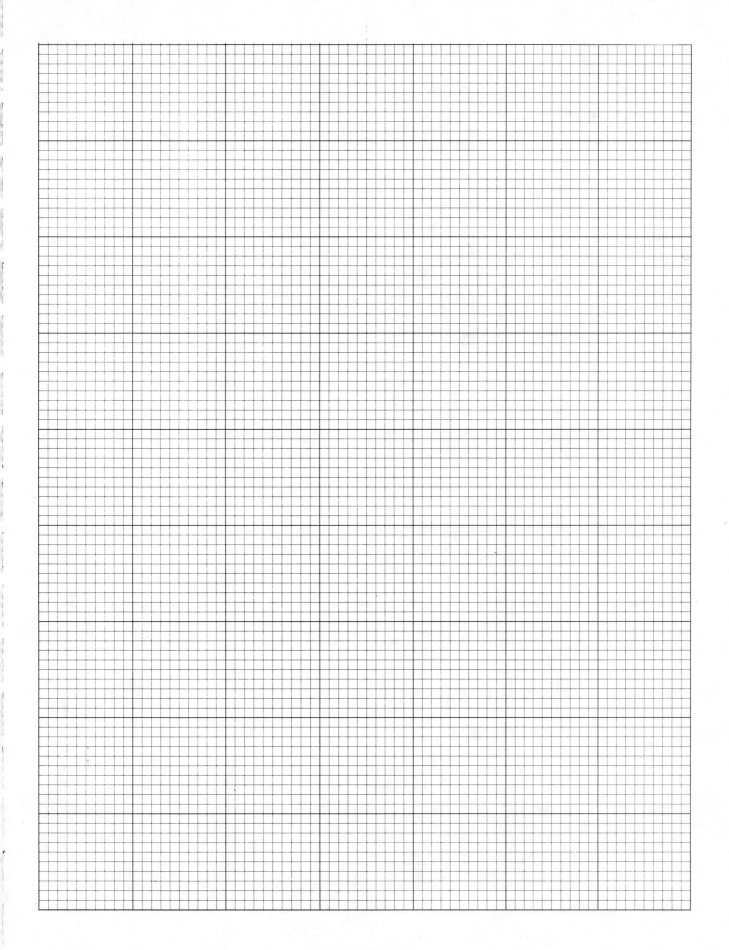

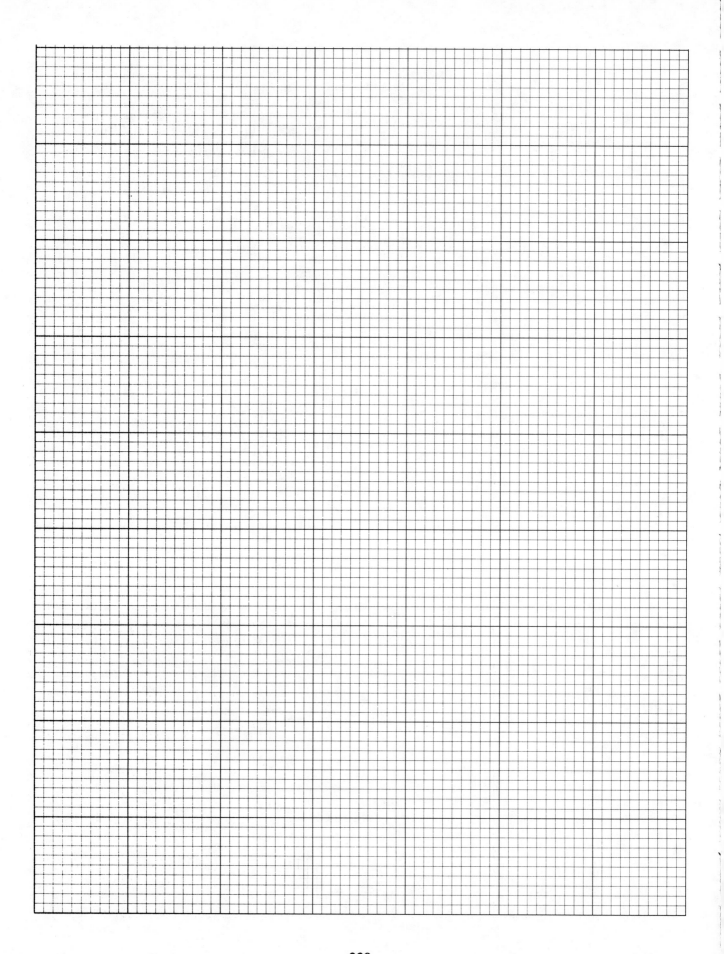

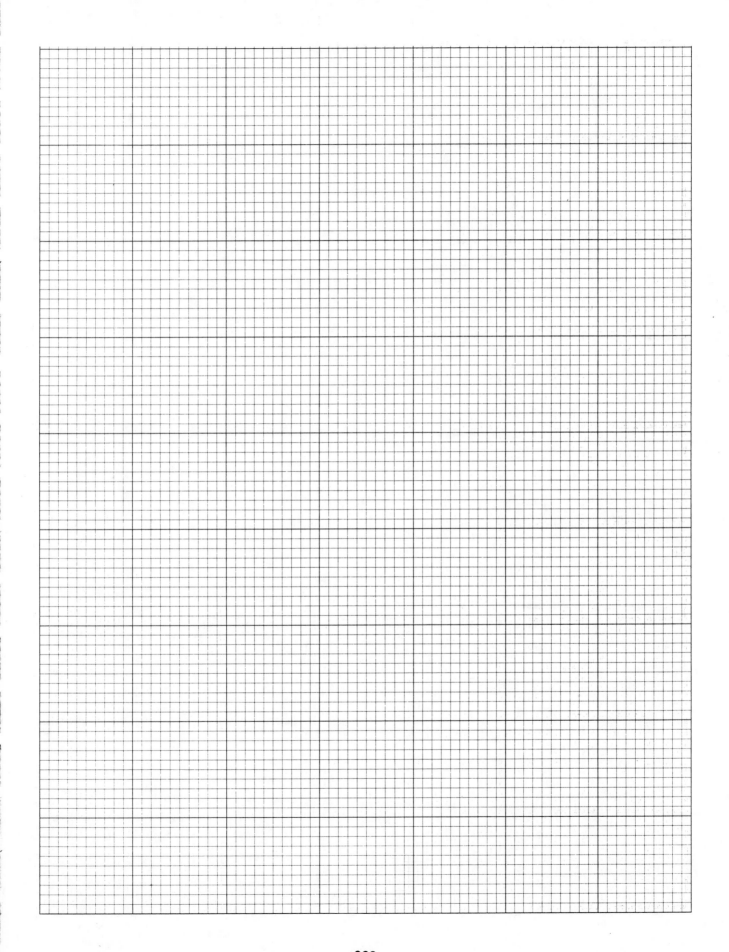

239

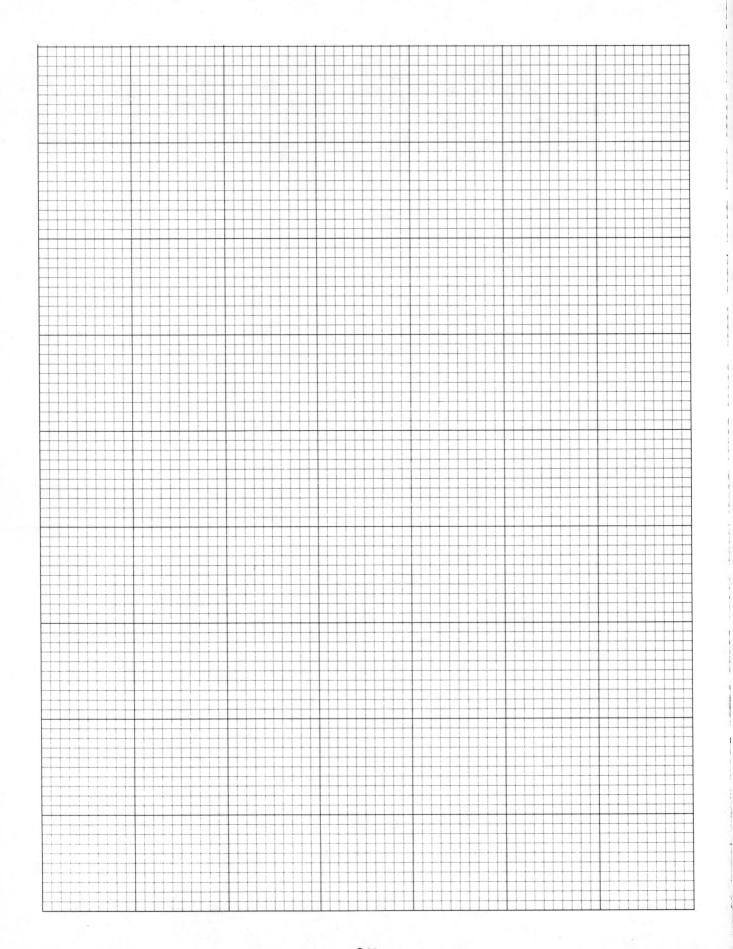